METHODS IN MOLECULAR BIOLOGY

For further volumes:
http://www.springer.com/series/7651

Chromatin Immunoprecipitation

Methods and Protocols

Edited by

Neus Visa and Antonio Jordán-Pla

Department of Molecular Biosciences, The Wenner-Gren Institute
Stockholm University, Stockholm, Sweden

Humana Press

Editors
Neus Visa
Department of Molecular Biosciences
The Wenner-Gren Institute
Stockholm University
Stockholm, Sweden

Antonio Jordán-Pla
Department of Molecular Biosciences
The Wenner-Gren Institute
Stockholm University
Stockholm, Sweden

ISSN 1064-3745 ISSN 1940-6029 (electronic)
Methods in Molecular Biology
ISBN 978-1-4939-7379-8 ISBN 978-1-4939-7380-4 (eBook)
DOI 10.1007/978-1-4939-7380-4

Library of Congress Control Number: 2017953868

Printed on acid-free paper

This Humana Press imprint is published by Springer Nature
The registered company is Springer Science+Business Media LLC
The registered company address is: 233 Spring Street, New York, NY 10013, U.S.A.

Preface

Chromatin immunoprecipitation (ChIP) is a powerful technique used to detect, map, and quantify interactions between a protein of interest and genomic DNA in vivo. In a typical ChIP experiment, cells or tissues are treated with formaldehyde to generate reversible crosslinks between DNA and other molecules located in the chromatin, typically proteins but also RNAs. The crosslinked chromatin is extracted and fragmented, and the resulting material is immunoprecipitated with antibodies against a protein of interest. The immunoprecipitated DNA is finally purified and analyzed. When ChIP is used for studies of specific genomic regions, the analysis of the immunoprecipitated DNA is usually based on quantitative PCR. Instead of PCR, ChIP can be combined with microarray hybridization or parallel sequencing, in which case ChIP provides occupancy maps for the protein of interest on a genome-wide basis.

Excellent ChIP protocols can be found in the literature (*see* for example reference 1 and references therein) and on websites of companies specialized in the commercialization of ChIP antibodies and reagents [2]. Guidelines for ChIP have also been made available by ENCODE and modENCODE consortia [3]. A specialized volume on ChIP protocols with focus on analysis of DNA methylation and on protocols for chromatin conformation capture assays was edited by P. Collas and published in this same series in 2009 [4]. However, ChIP assays are constantly evolving and being adapted to specialized applications, and this present volume extends the previous edition and includes protocols that illustrate the broad use of ChIP and ChIP-related methods in a variety of biological research areas. The volume also includes protocols that improve the performance of ChIP for specific applications. We hope that this volume of *Methods in Molecular Biology* on ChIP and ChIP-related techniques, including bioinformatic analysis of ChIP data, will be of interest to a very broad research community in the fields of biochemistry, molecular biology, microbiology, and biomedicine.

Stockholm, Sweden *Neus Visa*
Antonio Jordán-Pla

References

1. Carey MF, Peterson CL, Smale ST (2009) Chromatin Immunoprecipitation (ChIP) Cold Spring Harb Protoc. 2009(9):pdb.prot5279. doi: 10.1101/pdb.prot5279
2. http://www.abcam.com/index.html?pageconfig=popular_protocols
3. Landt SG, Marinov GK, Kundaje A et al (2012) ChIP-seq guidelines and practices of the ENCODE and modENCODE consortia. Genome Res 22:1813–1831
4. Collas P (ed) (2009) Chromatin immunoprecipitation assays. Methods in molecular biology 567. doi: 10.1007/978-1-60327-414-2_1

Contents

Contributors

SOPHIE H. ADJALLEY • *Wellcome Trust Sanger Institute, Hinxton, UK*
PATRIK ASP • *Albert Einstein College of Medicine, Michael F. Price Center, Bronx, NY, USA*
RICHÁRD BÁRTFAI • *Department of Molecular Biology, Radboud Institute for Molecular Life Sciences, Nijmegen, The Netherlands*
TIMOTHY V. BEISCHLAG • *Faculty of Health Sciences, Simon Fraser University, Burnaby, BC, Canada*
THOMAS CARROLL • *Medical Research Council, London Institute of Medical Sciences, Imperial College London, London, UK*
DIOGO S. CASTRO • *Instituto Gulbenkian de Ciência, Oeiras, Portugal*
CHRISTOPHE D. CHABBERT • *Genome Biology Unit, European Molecular Biology Laboratory (EMBL), Heidelberg, Germany; Institute of Molecular Health Sciences, ETH Zurich, Zürich, Switzerland*
ANDREA B. EBERLE • *Department of Chemistry and Biochemistry, University of Bern, Bern, Switzerland*
ANN-KRISTIN ÖSTLUND FARRANTS • *Department of Molecular Biosciences, The Wenner-Gren Institute, Stockholm University, Stockholm, Sweden*
HUI GAO • *Department of Biosciences and Nutrition, Novum, Karolinska Institutet, Huddinge, Sweden*
GREGOR D. GILFILLAN • *Department of Medical Genetics, Oslo University Hospital and University of Oslo, Oslo, Norway*
OLIVER HANKINSON • *Department of Pathology and Laboratory Medicine, University of California, Los Angeles, Los Angeles, CA, USA*
BRETT R. HANSON • *Roche Molecular Diagnostics, Los Gatos, CA, USA*
WIETEKE ANNA MARIA HOEIJMAKERS • *Department of Molecular Biology, Radboud Institute for Molecular Life Sciences, Nijmegen, The Netherlands*
ANTONIO JORDÁN-PLA • *Department of Molecular Biosciences, The Wenner-Gren Institute, Stockholm University, Stockholm, Sweden*
CHANDRASEKHAR KANDURI • *Department of Medical Biochemistry and Cell Biology, Institute of Biomedicine, The Sahlgrenska Academy, University of Gothenburg, Gothenburg, Sweden*
ON SUN LAU • *Department of Biological Sciences, National University of Singapore, Singapore, Singapore*
ANN-CHRISTIN LINDÅS • *Department of Molecular Biosciences, The Stockholm University, Stockholm, Sweden*
TANMOY MONDAL • *Department of Medical Biochemistry and Cell Biology, Institute of Biomedicine, The Sahlgrenska Academy, University of Gothenburg, Gothenburg, Sweden*
VICENT PELECHANO • *SciLifeLab, Department of Microbiology, Tumor and Cell Biology, Karolinska Institutet, Solna, Sweden*
GRATIEN G. PREFONTAINE • *Faculty of Health Sciences, Simon Fraser University, Burnaby, BC, Canada*
XIAOJUN REN • *Department of Chemistry, University of Colorado Denver, Denver, CO, USA*
TEODORA RIBARSKA • *Molecular Biology of the Cell II, DKFZ, Heidelberg, Germany*

Amarjit Saini • *Department of Laboratory Medicine, Clinical Physiology Karolinska Institutet and Unit of Clinical Physiology, Karolinska University Hospital, Stockholm, Sweden*

Ines de Santiago • *Li Ka Shing Centre, Cancer Research UK Cambridge Institute, University of Cambridge, Cambridge, UK*

Mário A.F. Soares • *Instituto Gulbenkian de Ciência, Oeiras, Portugal*

Lars M. Steinmetz • *Genome Biology Unit, European Molecular Biology Laboratory (EMBL), Heidelberg, Germany; Genetics Department, Stanford Genome Technology Center, Stanford University School of Medicine, Palo Alto, CA, USA*

Santhilal Subhash • *Department of Medical Biochemistry and Cell Biology, Institute of Biomedicine, The Sahlgrenska Academy, University of Gothenburg, Gothenburg, Sweden*

Carl Johan Sundberg • *Department of Laboratory Medicine, Clinical Physiology Karolinska Institutet and Unit of Clinical Physiology, Karolinska University Hospital, Stockholm, Sweden; Molecular Exercise Physiology, Department of Physiology and Pharmacology, Karolinska Institutet, Stockholm, Sweden; Department of Learning, Informatics, Management and Ethics, Karolinska Institutet, Stockholm, Sweden*

Ming Tan • *Medical Sciences I, Department of Microbiology and Molecular Genetics, University of California Irvine, Irvine, CA, USA*

Roubina Tatavosian • *Department of Chemistry, University of Colorado Denver, Denver, CO, USA*

Neus Visa • *Department of Molecular Biosciences, The Wenner-Gren Institute, Stockholm University, Stockholm, Sweden*

Kun Wang • *Department of Molecular Biosciences, The Stockholm University, Stockholm, Sweden*

Shenqi Wang • *Department of Biological Sciences, National University of Singapore, Singapore, Singapore*

Chunyan Zhao • *Department of Biosciences and Nutrition, Novum, Karolinska Institutet, Huddinge, Sweden*

Chapter 1

ChIP and ChIP-Related Techniques: Expanding the Fields of Application and Improving ChIP Performance

Neus Visa and Antonio Jordán-Pla

Abstract

Protein–DNA interactions in vivo can be detected and quantified by chromatin immunoprecipitation (ChIP). ChIP has been instrumental for the advancement of epigenetics and has set the groundwork for the development of a number of ChIP-related techniques that have provided valuable information about the organization and function of genomes. Here, we provide an introduction to ChIP and discuss the applications of ChIP in different research areas. We also review some of the strategies that have been devised to improve ChIP performance.

Key words Protein–nucleic acid interactions, Crosslinking, Immunochemistry, Polymerase chain reaction, Next-generation sequencing, Epigenetics, Genomics, Cell nucleus, Histones

1 Introduction to Chromatin Immunoprecipitation (ChIP)

Immunochemistry methods in general, and immunoprecipitation in particular, have been instrumental for the advancement of many research areas in molecular biosciences and biomedicine. Chromatin immunoprecipitation (ChIP) is a method that has been extensively used to study protein–DNA interactions in vivo. ChIP provides information about the association of a protein of interest, for example, a transcription factor or a post-translationally modified histone, with specific sequences in the DNA (*candidate sequence analysis*). ChIP can also be used to map the distribution of the protein of interest throughout the genome (*genome-wide analysis*). Regardless of the specific aim of the experiment, ChIP protocols involve the following basic steps:

1. Protein–DNA crosslinking in vivo, which is typically achieved by treating the biological material (cells in culture, tissues, or organs) with formaldehyde.
2. Extraction and fragmentation of the chromatin, which involves either sonication or enzymatic digestion of the fixed chromatin.

Neus Visa and Antonio Jordán-Pla (eds.), *Chromatin Immunoprecipitation: Methods and Protocols*, Methods in Molecular Biology, vol. 1689, DOI 10.1007/978-1-4939-7380-4_1, © Springer Science+Business Media LLC 2018

3. Immunoprecipitation of chromatin fragments that contain the protein of interest using a specific antibody against the protein (or against a post-translationally modified form of it).
4. Purification and analysis of the immunoprecipitated DNA. In a candidate sequence approach, the standard readout method is polymerase chain reaction (PCR), either end-point PCR (semi-quantitative) or quantitative real-time PCR (qPCR), whereas the genome-wide studies usually involved the use of microarray hybridization or massive parallel sequencing.

ChIP is nowadays a method accessible to any molecular biology laboratory, although its proper application is not trivial. Chapter 2 in this issue reviews the main steps of the method and provides technical considerations of general interest. Chapter 3 provides general guidelines for the quantification of ChIP results by qPCR. Chapter 4 presents a ChIP protocol combined with genome-wide profiling by microarray hybridization (ChIP-on-chip) and Chapter 5 describes the typical ChIP-seq method in detail. The specific conditions for extraction of chromatin can vary depending on the biological material under study. Chapters 5, 11, 12, and 15 present protocols that have been successfully applied to different types of samples, including human and plant tissues, cultured cells, and microorganisms.

2 The Origin and Diversification of ChIP

Most of the early experiments involving immunoprecipitation of proteins from chromatin preparations were based on the use of native chromatin and were focused on the analysis of the immunoprecipitated protein moieties. For example, Torres-Martinez and Ruiz-Carrillo [1] carried out experiments in which antibodies against histone H5 were used to immunoprecipitate mononucleosomes generated by digesting chromatin extracted from hen erythrocytes with micrococcal nuclease (MNase). The presence of different linker histones in the isolates was analyzed, and the results provided information about the exchangeability of linker histones [1]. The origin of ChIP dates back to the mid-1980s and was marked by the introduction of crosslinking treatments to stabilize protein–DNA interactions in experiments aimed at determining interactions between a protein of interest and chromosomal DNA sequences in vivo, an idea that was first launched by Lis and coworkers [2]. In their initial protocol, designed to study the interaction of RNA polymerase with the promoter of the *lac* gene in *Escherichia coli*, Gilmour and Lis used a UV source to irradiate bacterial cultures and in this way crosslink proteins to DNA in vivo. The crosslinked protein–DNA adducts were fragmented by sonication and used in immunoprecipitation experiments, and the immunoprecipitated DNA was finally analyzed by dot blot hybridization. A similar approach was applied soon thereafter to study the interaction of RNA polymerase II with genes of *Drosophila melanogaster* [3]. In this second report, the use of restriction enzymes was shown to be a viable alternative for chromatin fractionation, and the analysis of the immunoprecipitated DNA was extended to Southern blotting [3]. The broad applicability of the method was already pointed out in these pioneering studies, which identified the two basic premises for the use of ChIP: that antibodies are available against the protein of interest, and that the protein can be crosslinked to DNA in vivo at detectable levels.

The use of UV irradiation was soon replaced by chemical fixation with formaldehyde for reversible protein–DNA crosslinking [4], and polymerase chain reaction (PCR) became the preferred method for qualitative and quantitative analyses of immunoprecipitated DNA.

The use of ChIP gained particular attention as a result of its impact in the field of epigenetics, a field in which the combination of ChIP with genome-wide profiling technologies has provided fundamental knowledge on the molecular mechanisms that regulate chromatin structure and gene expression in eukaryotic cells. Because of the extensive use of ChIP in epigenetics, ChIP has often been regarded as a method for studies of histone modifications. The applications of ChIP are though much broader because ChIP can detect interactions between DNA and any protein of interest provided that the protein can be crosslinked, directly or indirectly, to the DNA in vivo. As a general rule, proteins that are in direct contact with DNA, such as histones, are efficiently crosslinked to chromosomal DNA and are relatively easy to analyze by ChIP. However, the generic nature of the crosslinker agents used in ChIP, usually formaldehyde, and the frequent occurrence of multiple crosslinking events in the conditions of a typical ChIP experiment, make it also possible to use ChIP for studies of proteins that associate with the DNA indirectly. Furthermore, a considerable fraction of chromatin consists of RNAs and RNA-binding proteins [5], and ChIP-related assays have been developed to analyze these components in a chromatin context. For example, ChIP has been used to demonstrate the co-transcriptional recruitment of RNA processing factors and decay factors to transcribed genes [*see* for example references 6–8]. And chromatin-RNA immunoprecipitation protocols (ChRIP) make it possible to isolate and characterize RNAs attached to chromatin [9]. In Chapter 6, Kanduri and coworkers describe an improved ChRIP protocol that is suitable to profile chromatin-associated RNA transcripts from different chromatin compartments across the genome.

Not surprisingly, methods for measuring chromatin packaging or determining the position of nucleosomes in the genome have attracted much interest. Nucleases have been used as tools to probeMicroarray chromatin structure since the 1970s [*see* for example references 10 and 11] and micrococcal nuclease (MNase) is currently used in many ChIP protocols to fragment fixed chromatin (*see* for example the Bar-ChIP protocol presented in Chapter 16). MNase was initially isolated from *Staphylococcus aureus* and can catalyze the cleavage of RNA and DNA to 3′-phosphonucleotide end-products. This enzyme has exo- and endo-ribonucleolytic activities, and can cleave both single- and double-stranded nucleic acids. The reason why MNase has become a powerful tool for chromatin mapping is that it is possible to find conditions in which the enzyme discriminates between nucleosome linker regions and DNA sequences that are engaged in nucleosome packaging. As a result, MNase can preferentially cleave the linkers and render free mononucleosomes [12]. Many protocols build on the use of MNase in combination with different readout methods, and excellent MNase protocols have been published [*see* for example reference 13]. A detailed protocol for the use of MNase for quantitative assessments of chromatin accessibility at specific genomic loci (MNase-qPCR) is described by Östlund Farrants in Chapter 7, and a protocol for mapping the distribution of nucleosomes genome-wide (MNase-seq) is provided in Chapter 8 by Hoeijmakers and Bártfai.

The characterization of multi-protein complexes in chromatin has also attracted much attention. In Chapter 9, Beischlag and coworkers present a protocol for sequential ChIP, a method for co-occupancy studies in which two successive rounds of immunoprecipitation are carried out to determine whether two proteins of interest coexist in the same chromatin region. A related question that has been particularly challenging is the determination of the stoichiometry of protein complexes that are associated with chromatin. This question has recently been resolved by single-molecule ChIP imaging (Sm-ChIPi), a method that combines a modified version of ChIP with single-molecule fluorescence imaging (*see* Chapter 10 by Tatavosian and Ren).

The original ChIP protocols used cultured cells as a starting material. The analysis of chromatin from tissues and organs is generally more challenging, but extremely interesting for developmental biologists and physiologists. Protocol modifications to analyze DNA–protein interactions in a physiological context have been reported and have provided satisfactory results. For example, Saini and Sundberg describe a protocol for ChIP that has been tailored for human muscle samples and has been successfully applied in investigations of epigenetic regulation of skeletal muscle (*see* Chapter 11). ChIP protocols have also been adapted to organs from model organisms, for example, fruitflies [14] or frogs [15], and to different types of microorganisms including extremophile organisms such as archaea (*see* Chapter 12 by Wang and Lindås). Interesting examples of ChIP specialization have also emerged from the field of infection biology, where ChIP is used for the study of host–microbe interactions. As an example of this type of application, Hanson and Tan describe in Chapter 13 a specialized ChIP assay called Intra-ChIP to study the association of microbial transcription factors with their target DNA sequences in the genome of infected eukaryotic cells.

3 Improving the Performance of ChIP Methods

The ChIP protocols have not only been adapted to a broad range of applications and biological materials, but have also been developed in different ways to overcome technical limitations of ChIP. Many of the chapters included in this volume address technical aspects of the ChIP technique and propose strategies that aim at improving its performance.

As in other biotechnology areas, the forefront of ChIP development aims at profiling the chromatin of individual cells, which requires a combination of microfluidics, DNA barcoding, and sequencing [16–18]. Single-cell ChIP-seq is still under development and is limited by the requirement of custom-made microfluidic devices, by the small amount of sequence data obtained, and by the insufficient sensitivity of the method. Nevertheless, the sensitivity of the ChIP-seq technique can be improved, and protocols for ChIP analysis of low cells numbers have been successfully applied. One strategy utilized to increase ChIP sensitivity is the use of native, non-crosslinked chromatin. Native chromatin offers higher epitope accessibility than fixed chromatin, which improves the efficiency of the immunoprecipitation and makes it possible to obtain consistent results with lower cell numbers. In Chapter 14, Ribarska and Gilfillan present a Native-ChIP procedure based on the use of non-crosslinked chromatin from fresh and live-frozen cells that works with as few as 4×10^4 cells. In this Native-ChIP method, the native chromatin is fractionated with micrococcal nuclease instead of sonication.

The level of expression of the protein of interest is another parameter that imposes demands on the sensitivity of ChIP, as low abundance proteins often result in poor ChIP signals. Wang and Lau present in Chapter 15 a modified ChIP protocol called Maximized Objects for Better Enrichment (MOBE)-ChIP, which significantly improves ChIP sensitivity and allows cell type-specific studies of low-abundance proteins. MOBE-ChIP uses crosslinked chromatin and consists in scaling up the ChIP experiment and in using high-affinity antibodies, which increases the signal-to-noise ratio.

Classical ChIP methods interrogate one protein at a time, but some experiments may require the analysis of several proteins, for example, different histone modifications, in the same set of samples. In other cases, the experiment may comprise multiple treatments, which increases the number of samples to be analyzed in parallel. ChIP experiments with large numbers of samples are time-consuming and highly laborious in the absence of robotic instrumentation. Barcoded high-throughput ChIP-Seq (Bar-ChIP) addresses this issue and allows the simultaneous processing of multiple ChIP-seq samples. The strategy of Bar-ChIP consists in the direct ligation of unique molecular barcodes to the fractionated chromatin extracted from each sample prior to immunoprecipitation. Following this approach, the DNA fragments from different samples can be pooled together and subjected to immunoprecipitation because each sample carries a sample-specific tag. The tags are adaptors that are compatible with Illumina sequencing. The adaptor sequences allow the sorting of reads from individual samples in silico. In this way, Bar-ChIP enables the efficient and simultaneous profiling of multiple samples (*see* reference 19 and Chapter 16 in this volume).

The computational analysis of genome-wide ChIP data requires the use of specialized bioinformatics tools and constitutes a challenging task for many research laboratories. In Chapter 17, De Santiago and Caroll present an overview of the methods that are currently available for the analysis of ChIP-seq data. The same chapter provides a detailed description of the computational procedures that are routinely used for the analysis of ChIP-seq data in R/Bioconductor, including instructions from which readers can construct their own analysis pipelines.

The use of standards for sample-to-sample normalization is important in ChIP experiments when different samples, for example, controls and treated samples, have to be compared to each other. In ChIP-qPCR experiments, an invariable genomic region can be chosen as a reference, analyzed in parallel and used for normalization (internal standard). However, the effect of experimental treatments on the chromatin is often unpredictable and this lack of knowledge often precludes the choice of a suitable reference region. Instead of using an internal reference, a possible solution is to spike in chromatin from another species that can be distinguished in silico from the chromatin of the organism of interest. The amount of exogenous chromatin immunoprecipitated by the antibody can then be used to normalize data from different samples [20, 21]. This strategy works well in experiments with antibodies against conserved proteins, for example, histones, that recognize the protein of interest in both species. However, ChIP experiments with antibodies against less conserved antigens cannot take advantage of this method. An alternative, more universal strategy is the use of a synthetic external reference. The production and use of one such external standard is described by Eberle in Chapter 18. In this case, a synthetic DNA sequence is labeled with digoxygenin (DIG) and crosslinked to an anti-DIG antibody. The antibody–DIG–DNA complex is then used as external spike-in reagent [22].

Acknowledgment

This work was financed by grants from The Swedish Research Council and The Swedish Cancer Society to N.V. A.J.P. was supported by the Department of Molecular Biosciences, The Wenner-Gren Institute at the Stockholm University.

References

1. Torres-Martinez S, Ruiz-Carrillo A (1982) Nucleosomes containing histones H1 or H5 are closely interspersed in chromatin. Nucleic Acids Res 10:2323–2335
2. Gilmour DS, Lis JT (1984) Detecting protein-DNA interactions in vivo: distribution of RNA polymerase on specific bacterial genes. Proc Natl Acad Sci U S A 81:4275–4279
3. Gilmour DS, Lis J (1985) In vivo interactions of RNA polymerase II with genes of Drosophila melanogaster. Mol Cell Biol 5:2009–2018
4. Solomon MJ, Larsen PL, Varshavsky A (1988) Mapping protein-DNA interactions in vivo with formaldehyde: evidence that histone H4 is retained on a highly transcribed gene. Cell 53:937–947
5. Paul J, Duerksen JD (1975) Chromatin-associated RNA content of heterochromatin and euchromatin. Mol Cell Biochem 9:9–16
6. Kotovic KM, Lockshon D, Boric L, Neugebauer KM (2003) Cotranscriptional recruitment of the U1 snRNP to intron-containing genes in yeast. Mol Cell Biol 23:5768–5779
7. Morlando M, Ballarino M, Gromak N et al (2008) Primary microRNA transcripts are processed co-transcriptionally. Nat Struct Mol Biol 15:902–909
8. Hessle V, Björk P, Sokolowski M et al (2009) The exosome associates cotranscriptionally with the nascent pre-mRNP through interactions with heterogeneous nuclear ribonucleoproteins. Mol Biol Cell 20:3459–3470
9. Listerman I, Sapra AK, Neugebauer KM (2006) Cotranscriptional coupling of splicing factor recruitment and precursor messenger RNA splicing in mammalian cells. Nat Struct Mol Biol 13:815–822
10. Rill R, Van Holde KE (1973) Properties of nuclease-resistant fragments of calf thymus chromatin. J Biol Chem 248:1080–1083
11. Axel R, Cedar H, Felsenfield G (1975) The structure of the globin genes in chromatin. Biochemistry 14:2489–2495
12. Hewish DR, Burgoyne LA (1973) Chromatin sub-structure. The digestion of chromatin DNA at regularly space sites by a nuclear deoxyri-bonuclease. Biochem Biophys Res Commun 52:504–510
13. Zaret K (2005) Micrococcal nuclease analysis of chromatin structure. Curr Protoc Mol Biol 21:1. doi:10.1002/0471142727.mb2101s69
14. Legube G, McWeeney SK, Lercher MJ, Akhtar A (2006) X-chromosome-wide profiling of MSL-1 distribution and dosage compensation in Drosophila. Genes Dev 20:871–878
15. Stewart D, Tomita A, Shi YB, Wong J (2006) Chromatin immunoprecipitation for studying transcriptional regulation in Xenopus oocytes and tadpoles. Methods Mol Biol 322:165–181
16. Dahl JA, Collas P (2009) A rapid micro chromatin immunoprecipitation assay (micro-ChIP). Nat Protoc 3:1032–1045
17. Gilfillan GD, Hughes T, Sheng Y et al (2012) Limitations and possibilities of low cell number ChIP-seq. BMC Genomics 13:645. doi:10.1186/1471-2164-13-645
18. Rotem A, Ram O, Shoresh N et al (2015) Single-cell ChIP-seq reveals cell subpopulations defined by chromatin state. Nat Biotechnol 33:1165–1172
19. Chabbert CD, Adjalley SH, Klaus B et al (2015) A high-throughput ChIP-Seq for large-scale chromatin studies. Mol Syst Biol 11:777. doi:10.15252/msb.20145776
20. Bonhoure N, Bounova G, Bernasconi D et al (2014) Quantifying ChIP-seq data: a spiking method providing an internal reference for sample-to-sample normalization. Genome Res 24:1157–1168
21. Orlando DA, Chen MW, Brown VE et al (2014) Quantitative ChIP-Seq normalization reveals global modulation of the epigenome. Cell Rep 9:1163–1170
22. Eberle AB, Böhm S, Östlund Farrants AK et al (2012) The use of a synthetic DNA-antibody complex as external reference for chromatin immunoprecipitation. Anal Biochem 426:147–152

Chapter 2

Considerations on Experimental Design and Data Analysis of Chromatin Immunoprecipitation Experiments

Antonio Jordán-Pla and Neus Visa

Abstract

Arguably one of the most valuable techniques to study chromatin organization, ChIP is the method of choice to map the contacts established between proteins and genomic DNA. Ever since its inception, more than 30 years ago, ChIP has been constantly evolving, improving, and expanding its capabilities and reach. Despite its widespread use by many laboratories across a wide variety of disciplines, ChIP assays can be sometimes challenging to design, and are often sensitive to variations in practical implementation.

In this chapter, we provide a general overview of the ChIP method and its most common variations, with a special focus on ChIP-seq. We try to address some of the most important aspects that need to be taken into account in order to design and perform experiments that generate the most reproducible, high-quality data. Some of the main topics covered include the use of properly characterized antibodies, alternatives to chromatin preparation, the need for proper controls, and some recommendations about ChIP-seq data analysis.

Key words ChIP, ChIP-seq, Guidelines, Quality control, Replicates, High-throughput sequencing, Standards, Antibodies, Genome-wide, Library

1 Introduction to Chromatin

The genomes of eukaryotic organisms are packed into chromatin. Chromatin helps condense the DNA into chromosomes that fit it inside the cell nucleus and allow cells to split their genomes during cell division. Chromatin is intimately involved in other processes like gene regulation, DNA repair, and DNA replication. Chromatin is a complex of DNA, RNA, and associated proteins [1]. The basic structural units of chromatin are called nucleosomes and comprehend 147 bp of DNA wrapped around an octamer of proteins called histones [2]. Certain residues in the amino (N)-terminal tails of histones can be chemically modified post-translationally. Among the many possible histone tail modifications, the most common are methylations and acetylations of lysine residues [3]. Some histone modifications change the affinity of the histone tails for DNA,

Neus Visa and Antonio Jordán-Pla (eds.), *Chromatin Immunoprecipitation: Methods and Protocols*,
Methods in Molecular Biology, vol. 1689, DOI 10.1007/978-1-4939-7380-4_2, © Springer Science+Business Media LLC 2018

affecting the strength by which DNA wraps around nucleosomes. In a similar manner, the DNA itself can be chemically modified by methylation or hydroxymethylation which alters its chemical and structural properties [4]. Together, DNA and histone modifications are mechanisms to affect gene expression that are not codified in the DNA sequence and that can be transmitted through cell divisions. Such modifications are often referred to as epigenetic traits [5]. Chromatin can be found in two main forms: euchromatin and heterochromatin. Histone modifications can be established and reversed dynamically by different groups of enzymes, and modified histones serve as signals to recruit other DNA-binding proteins that can accomplish different tasks, such as moving and rearranging nucleosomes along the DNA strand, or initiating the assembly of the transcription pre-initiation complex.

2 What Is ChIP?

Complex and rich interactions between DNA and its associated RNAs and proteins govern essential cellular processes such as transcription, replication and DNA damage response, amongst many others. These processes have fascinated researchers for decades, and have been extensively studied. One of the techniques that has best helped dissect the way those interactions are established, paving the way for our understanding of their biological relevance, is the chromatin immunoprecipitation (ChIP). After initial implementations by Gilmour and Lis in 1984 and 1985 [6, 7], in which UV radiation was used to fix protein–DNA interactions in bacterial cells and *Drosophila melanogaster*, Solomon and collaborators modified the technique by replacing UV by formaldehyde as fixative [8]. This approach became the most widely adopted and is still broadly used nowadays, and the details of the ChIP protocol have been continuously modified and refined up to this day.

As its name indicates, ChIP relies on the use of antibodies to precipitate or pull-down a specific DNA-binding protein of interest and on the subsequent analysis of the DNA fragments attached, which makes it possible to discern the genomic loci the protein was bound to, in vivo. Protein–DNA interactions can be approached fundamentally in two ways: (a) by looking at the presence of specific sequences in the immunoprecipitated DNA (loci-specific analysis), or (b) by mapping all the loci a protein is bound to (genome-wide analysis).

One of the main strengths of ChIP relies on its immense versatility. It can be applied to the majority of organisms for which an antibody against a certain protein is available, and due to the vast array of antibodies that can be used to interrogate the genome for the presence of certain proteins and their modifications, it is possible to address very complex chromatin biology aspects. One of the questions that can be analyzed by ChIP is the differential binding patterns of certain proteins to specific regions, and in various situations: during the cell cycle, in response to internal or external cues, between different cell types, etc. Thanks to these types of studies, we have acquired a comprehensive view on how transcription factors, modified nucleosomes, RNA polymerases and other proteins interact with DNA, and with each other, in the chromatin context [9–11].

3 Experimental Design Considerations

Traditionally, ChIP protocols have been hands-on intensive, often taking 3–4 days to complete. They have also required large cell numbers as input material, and involved extensive sample handling. Large and cumbersome protocols are more susceptible to different types of errors that end up contributing to overall inconsistencies between replicates. To overcome these challenges and expand ChIP capabilities, researchers have devised different approaches to make ChIP experiments faster, simpler, less prone to human error, and compatible with lower input cell numbers.

Regardless of specific aims, the majority of protocols share the same basic principles. In a typical ChIP experiment (Fig. 1), cells are treated with formaldehyde to fix protein–DNA interactions occurring in vivo. Once these contacts are stabilized, cell membranes are disrupted to release the chromatin. Free chromatin is then fragmented by sonication, enzymatic digestion, or both. Once chromatin is fragmented, an antibody is used to capture specific protein–DNA complexes and the immunoprecipitated complexes are washed to eliminate unspecific interactions. Finally, the cross-linking is reversed, the proteins are removed and the DNA fragments are analyzed by different genetic techniques, such as PCR, microarray hybridization, or deep sequencing.

Each of the above steps can be implemented in different ways, depending on the particular needs of the experiment. However, some common aspects need to be taken into account when designing ChIP experiments in order to take full advantage of the versatility of ChIP without sacrificing reliability and reproducibility.

3.1 Use of Protease Inhibitors

Precautions need to be taken to avoid protein and DNA degradation during chromatin preparation and IP. For that purpose, the strategy is twofold: (1) add protease inhibitor cocktails to washing and extraction buffers throughout the protocol until final ChIP'ed DNA is recovered, and (2) keep the samples cold by placing them on ice during manipulations, centrifuging at 4 °C and performing antibody incubations and washes in a cold chamber.

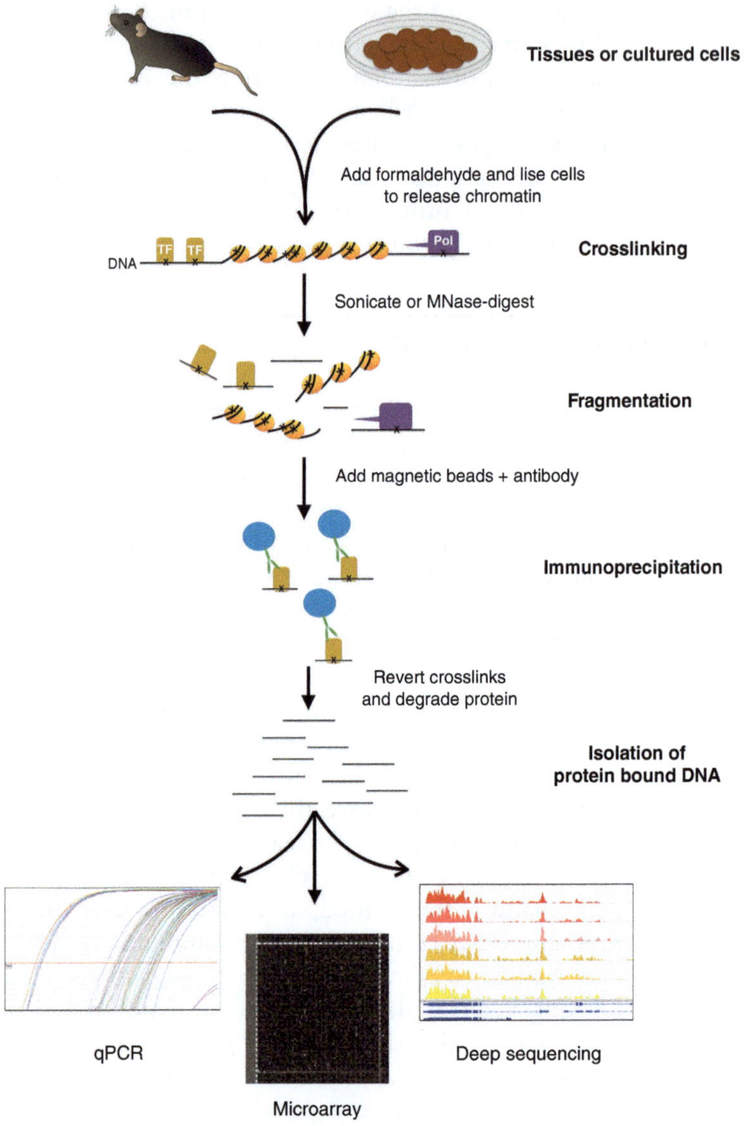

Fig. 1 Outline of a typical X-ChIP assay. Starting from cultured cells, homogenized tissues or other materials, formaldehyde is added to crosslink DNA to proteins (*black crosses*). Formaldehyde efficiently crosslinks DNA to transcription factors (TF), nucleosomes (*orange ovals*), RNA polymerases (Pol), or other chromatin proteins. Chromatin is then fragmented into small pieces by either sonication or Micrococcal Nuclease (MNase) digestion. Antibody-conjugated magnetic beads are used to capture a specific type of protein–DNA complex (in this case TF-DNA). Once enriched, the DNA bound by the TF is released by reversal of crosslinking and protein degradation. DNA fragments can be analyzed by different genetic methodologies such as quantitative PCR, microarray hybridization, or deep sequencing

3.2 Input Cell Numbers

One of the challenges of the ChIP method is to obtain a good enough signal-to-noise (SN) ratio. Typical ChIP experiments start with 1–10 million cells. A good SN ratio with one million cells can be achieved for immunoprecipitations (IPs) against medium to highly abundant chromatin factors, such as unmodified histones or RNA polymerases. To obtain good SN ratios for less abundant proteins one might need to start with ten million cells instead [12]. Another factor that contributes to the success of the IP is antibody quality. A high-affinity antibody will allow for a reduction of cell number without increasing the noise. It is therefore important to experimentally determine the appropriate number of cells needed to obtain high SN ratios for a given protein of interest in the condition of the experiment. Recent advances that tackle cell number limitations are briefly commented in Subheading 6 of this chapter.

3.3 Replication

ChIP is a multi-step and often multi-day protocol in which many factors such as cell-culture conditions, sonication, and antibody batches might contribute to certain variability between experiments. It is therefore recommended to carry out experiments with biological replicates to increase the reliability and robustness of the generated data. Although there is no real consensus as to how many replicates should be performed, it is almost universally accepted that at least biological duplicates are necessary to have a certain statistical power that supports the potential biological findings [13]. In addition, and especially when performing high-throughput ChIP (ChIP-chip or ChIP-seq), it is recommended that the processing of all samples is made simultaneously to avoid introducing batch effects. In case this is not possible, one must make sure to include all replicates in the same batch, to be able to correct any batch effect, bioinformatically [14].

3.4 Crosslinking Options

The aim of crosslinking is to retain the protein of interest on its in vivo genomic binding location for further study. DNA–protein interactions that occur *in vivo* can be stabilized either by using chemical crosslinkers [15], or by exposing cells to UV radiation [16, 17]. An alternative approach is to analyze protein–DNA interactions without crosslinking. These approaches are known as crosslinked ChIP (X-ChIP) and Native ChIP (N-ChIP), respectively. When opting for X-ChIP, the most common crosslinking agent is formaldehyde (FA). Its small size makes it well suited for capturing interactions between macromolecules that are in close proximity (~2 Å apart) [18, 19]. Moreover, FA bonds can be reversed, whereas UV-induced bonds are permanent. Reversibility greatly allows for an efficient isolation of the immunoprecipitated DNA in the final steps of the protocol [20]. When opting for X-ChIP, the researcher has to pay attention not to over-crosslink the samples. Longer incubations with the fixative lead to excessive crosslinking, which in turn might lead to changes in epitope accessibility, difficulties in chromatin sonication and overall reduced immunoprecipitation efficiencies [21]. To avoid these negative effects, it is advisable to carry out time-course experiments in which samples are fixed with FA at room temperature during 2–30 min, sonicated, and the size of the DNA fragments inspected by agarose gel electrophoresis. A good fixation time is the one that yields DNA fragments with lengths between 200 and 600 bp [22]. It is recommended to determine the optimal fixation conditions for all different epitopes and cell lines used, as fixation times may vary greatly among them.

3.5 Native Chromatin Preparation

Crosslinking is sometimes required to fix weak or transient protein–DNA contacts, but certain proteins are very stably bound to the DNA *per se*, e.g., histones and RNA polymerases [23]. In such cases, it might be better to avoid using FA not to compromise epitope recognition by the antibodies. An additional advantage of preparing native chromatin is that the amount of recovered DNA after IP is higher. This is especially interesting when starting from small cell numbers (discussed further in Subheading 6). A pitfall of using native chromatin is that chromatin factors might rearrange along the DNA during sonication or enzymatic digestion of the chromatin. In this case, suitable controls may need to be used to account for such changes. Table 1 summarizes the pros and cons of choosing one method over the other.

3.6 Sonication or Enzymatic Digestion

Once the chromatin, either fixed or native, has been released from the nucleus, it has to be fragmented. Fragmentation facilitates antibody pulldown and increases the resolution of the mapping of protein-binding sites. The two most common methods of chromatin fragmenting are sonication and nucleolytic cleavage with micrococcal nuclease (MNase). X-ChIP protocols usually rely only on sonication, as crosslinked chromatin is more difficult to digest than native chromatin [24]. Sonication is typically achieved by applying a variable number of sonic pulses to the sample until the DNA is fragmented to a size that is suitable for ChIP (typically 200–600 bp, although this might vary depending on downstream applications). Care needs to be taken to ensure that the sonication does not result in degradation of the sample by overheating. To this end, modern sonicators are equipped with cooling units that keep the sample cold during the process. The alternative to DNA shearing by sonication is MNase digestion. MNase is a nuclease that cleaves naked dsDNA [25]. On average, it generates smaller fragments than sonication, which improves the resolution of the method. However, it has been shown that MNase does not cleave DNA randomly, having a preference for binding and cleaving specific sequences [26]. MNase digestion is gaining momentum within the ChIP community due to its increased solubilization capabilities and its compatibility with both X-ChIP and N-ChIP. Other methods have been developed that take advantage of both chromatin preparation strategies, applying sonication and enzymatic digestion sequentially to the same sample. This combination works especially well for

Table 1
Comparison between the two main strategies for chromatin preparation

	N-ChIP	X-ChIP
Advantages	Improved native epitope recognition	Good for weak or transient protein–DNA contacts
	Efficient precipitation of DNA and protein	Reduces rearrangements of nucleosomes or other factors during sonication/digestion
	Shorter fragments (higher resolution)	Applicable to the majority of organisms
	Compatible with low starting cell numbers	
Disadvantages	Inefficient capture of weak DNA binders	Epitope alterations caused by exposure to FA might lead to suboptimal antibody detection
	Enzymatic cleavage does not produce random fragments of chromatin	Loss of resolution as crosslinked chromatin is less susceptible to shearing by sonication
	Proteins might move along the DNA during chromatin preparation	Impaired accessibility to enzymes like MNase

tightly packed heterochromatic regions [27], and it is on the basis of the High-Resolution ChIP [28] and ChIP-exo [29] methodologies (*see* Subheading 5.9).

3.7 Immunoprecipitation and Antibodies

The quality of the antibody used is crucial to achieve specific IPs that yield good SN ratios and that are robust and reproducible. Some aspects to consider when choosing antibodies for ChIP are:

1. Polyclonal vs. Monoclonal. Polyclonal antibodies tend to perform better overall in ChIP because they recognize different epitopes of the same protein, which sometimes is useful to compensate for the potential loss of recognition sites caused by FA crosslinking. However, it has to be taken into account that monoclonal antibodies show less batch-to-batch variation compared to polyclonal antibodies.
2. Specificity and quality. Antibodies should be specific and efficient in recognizing the protein of interest over any other protein present in the chromatin fraction. They should preferentially be affinity-purified, although sera have also been used. In any case, a general recommendation is to test the antibodies in immunoblotting, immunofluorescence, and immunoprecipitation conditions [30]. Success in giving specific signal over background in these conditions is a good indication that an antibody will work in ChIP conditions. However, an antibody is considered ChIP grade only when it has been tested successfully in ChIP conditions.
3. Unavailability of ChIP-grade antibodies. In case there is no ChIP-grade antibody available for the target protein, but there is a good candidate that has been successful in all tests explained

in the last point, there are some further tests that can be performed to evaluate its performance in ChIP:

(a) To carry out an actual ChIP experiment and measure the enrichment of the protein of interest in positive and negative loci by PCR. Additionally, it is recommended to carry out ChIP in parallel with a known antibody (e. g. anti-histone H3) to ensure that the protocol used works as expected.

(b) In case the antibody works for ChIP, it is recommended to titrate its concentration to increase the SN ratio. To this end, test different antibody concentrations in a range of 1–10 μg for every 25 μg of chromatin.

(c) Elute the proteins bound to the beads after the last wash of the IP step and run a Western blotting with the eluate to see if the target protein was efficiently immunoprecipitated.

4. Use of antibodies against an epitope tag. In case of experiments that involve cell transfection with an epitope-tagged version of the target protein, the range of expression of the construct should remain in the physiological range, as close as possible to the real expression level of the endogenous target protein. In such scenarios, it is recommended to perform a parallel pull-down using the antibody against a tag but with chromatin prepared from cells that do not express the epitope-tagged protein.

5. Special case of histone modification antibodies. This type of antibody needs more validation in order to be safely considered as specific for the modification they are targeted for [31]. Taking as an example the histone modification H3K27me, proper antibodies should: (a) not recognize off-target modified histones, like H3K36me; (b) have low affinity for different methylation states of the same residue (e.g., H3K27me2, H3K27me3), and (c) be robust between different batches.

3.8 Control Samples

There are several factors that might contribute to artifacts during sample preparation. Non-uniform chromatin fragmentation during sonication, sequence bias in MNase digestions, and non-specific pulldowns are some of the main sources of variation that can challenge the acquisition of true biological signals [32]. The inclusion of appropriate controls is crucial to extract meaningful data out of a ChIP experiment. Some of the most important are:

(a) Input DNA: some regions of the genome are more accessible to digestion or sonication than others due to chromatin conformation. For example, euchromatic regions are more likely to be represented in the final ChIP'ed DNA than heterochromatic regions [33]. The same is true for promoter and terminator regions, which are often found as open sequences depleted of nucleosomes [34]. To account for these differences, it is recommended to extract DNA from an aliquot of crosslinked and sonicated chromatin before the IP step.

(b) Mock IPs: there are two main ways of measuring the levels of unspecific binding of an assay. The first is known as a "non-specific IP" control, and consists of using antibodies that should not recognize any particular protein in the extract. The most typically used non-specific antibody is IgG. An alternative is using a non-chromatin antibody-like anti-GFP. The second approach is to perform IP in the absence of any kind of antibody. This is what is called a "no-antibody" or "beads only" control, and allows measuring the background enrichment of DNA fragments produced simply by mixing chromatin with beads.

(c) Positive antibodies: the use of antibodies against histone modifications that signal the presence of active (H3K4me3) and silent (H3K9me3) promoters can be used to test that all steps of the ChIP protocol have been successfully performed. However, it is important to note that such type of controls requires prior knowledge about regions that either contain or not this type of modification. Additionally, in the case of histone modifications, they have to be always accompanied by a parallel pulldown against the unmodified histones (in this case H3) to account for potential histone rearrangements caused by chromatin sonication or digestion [35] (*see* Subheading 3.5).

(d) Spike-in controls: chromatin derived from remote organisms (e.g., yeast for human or fly ChIPs) can be added at known concentrations to the samples prior to IP to normalize IP'ed DNA internally against the levels of reference IP'ed DNA [36]. An alternative to external chromatin is the addition of custom-made antibody-DNA complexes. One example consists of anti-digoxigenin (anti-DIG) antibodies bound to DNA molecules of known sequence that have been previously labeled with DIG [37].

4 Data Collection After Immunoprecipitation

Once the DNA fraction that was bound to the target protein of interest has been recovered, there are different genetic methods that can be used to either quantify binding of the protein to particular genes or regions of interest, or to generate an unsupervised map of its binding to the whole genome. The most common ways of analyzing ChIP'ed DNA are: (a) PCR (ChIP-PCR), (b) array hybridization (ChIP-chip), and (c) deep sequencing (ChIP-seq).

(a) ChIP-PCR. It is used to measure levels of DNA at certain loci relative to the levels at other loci. The preferred method for this is real-time quantitative PCR (RT-qPCR). To account for technical biases and background binding, levels of DNA of a specific locus in the IP sample have to be normalized by the levels of DNA at the same locus on the matched control samples. On one hand, normalizing by Input DNA eliminates sonication or enzymatic digestion preferences and biases that might have occurred. On the other hand, dividing by the Mock IP samples informs about the strength of the signal over the background of the analysis. Care should be taken when designing primers for qPCR, paying special attention to aspects such as melting temperatures and GC content. Optimal concentrations should be calculated for each primer pair. To assess ChIP specificity, it is recommended to amplify positive and negative loci (where the target protein is known to be present or absent, respectively), as well as non-template controls (PCR reactions that contain everything but ChIP'ed DNA). RT-qPCR can be a quick and cost-effective method when the aim of the analysis is to investigate one or a few loci. Due to its high sensitivity and broad dynamic range, it is the preferred method for reaffirming and validating the findings derived from other techniques, like ChIP-chip and ChIP-seq.

(b) ChIP-chip. When the aim of the experiment is to acquire a genome-wide description of the binding sites of the protein of interest, a more suitable approach is to label the DNA and hybridize it to arrays [38]. Genome-wide methods have the advantage of being unsupervised and of allowing for the discovery of novel binding sites [39]. One of the most successful microarray technologies is Affymetrix tiling arrays [40], which allows for the mapping of binding sites at high resolution across the whole genome of various species. As is the case with qPCR, hybridization of Input, IgG, or unmodified histone controls are also needed to normalize and extract meaningful readouts.

(c) ChIP-seq. Similar to ChIP-chip, genome-wide binding profiles can be generated when the IP'ed DNA is directly sequenced with high-throughput sequencing (HTS) platforms. Sequenced reads are mapped to a reference genome and computationally analyzed to detect protein–DNA interactions. Genome-wide methods for the analysis of ChIP DNA (ChIP-chip and ChIP-seq) have improved dramatically over the last 10 years. They have become faster, simpler, more powerful and cheaper to use [41]. Table 2 compares different characteristics of both methodologies.

Table 2
Differences between ChIP-chip and ChIP-seq

	Coverage	Resolution	Noise sources	Dynamic range	Multiplexing	Cost
ChIP-chip	Whole genome (array- and species-specific)	5–35 bp (tiling)	Cross-hybridization and oligo sequence bias	Background noise at low hybridization levels and saturation at high levels	Not possible	~350$ per sample
ChIP-seq	Whole genome (depends on the annotation of the genome)	1 bp	Sequencing errors and GC biases	Broader	Possible	~250$ per sample (ten million reads)

5 ChIP-Seq Guidelines

In the last few years, ChIP-seq has become the gold standard for high-throughput ChIP assays [42]. A myriad of variants and refinements of the traditional ChIP method have been developed to study different aspects of the chromatin biology, with new approaches being constantly proposed. There are many deep sequencing platforms and technologies, each of which uses different library construction and sequencing reagents and workflows (for a review of NGS technologies, *see* Goodwin et al. [43]). Due to space constraints, we have chosen to focus in one of the most widely adopted platforms, Illumina. Here are some of the aspects that a researcher needs to take into account when designing ChIP-seq experiments.

5.1 Library Preparation

IP'ed DNA requires a series of processing steps before it is ready for sequencing. As mentioned, there are different protocols available depending on the NGS platform of choice. The most commonly used pipeline (Illumina-compatible) includes end-repair of DNA fragments, A-tailing, adaptor ligation at 5′ and 3′ ends, size selection, PCR amplification, and quality assessment with capillary electrophoresis systems (Bioanalyzer or similar). Altogether, these steps constitute what is known as "library preparation." One of the aspects that can negatively impact the quality of the library is the number of PCR cycles used to amplify the adaptor-ligated fragments [44]. Too much PCR template or too many PCR cycles might decrease specific signal and library complexity. It is therefore recommended to empirically determine the minimum number of PCR cycles needed to amplify the library.

5.2 Multiplexing

During library construction*Multiplex*, different samples can be labeled or barcoded with specific sequence tags. Once labeled, samples can be mixed and sequenced in a single lane of the sequencer to avoid lane-batch effects. Alternatively, chromatin can be barcoded after sonication and different samples containing different barcodes can be pooled and IP'ed together [45–47]. After sequencing, the samples can be easily de-multiplexed in silico. To increase library depth, one can run additional lanes and combine the output reads.

5.3 Sequencing

The individual fragments that constitute the ChIP-seq libraries can be sequenced from one or both ends of the DNA molecule. Reading the molecule from one end is known as single-end sequencing, whereas reading from both ends is called paired-end sequencing. Single-end reads with a length of approximately 50–75 nt are usually sufficient for ChIP analysis [48]. Paired-end reads can increase mappability, especially if the target protein of interest binds to repetitive parts of the genome, but their contribution is not nearly as important as it is for discriminating between alternative transcripts in RNA-seq or indels in DNA-seq. The number of reads required for accurate peak discovery depends on the type of protein factor under investigation. The patterns of protein binding are commonly classified into three types: narrow, broad, and mixed. The narrow type is usually seen in transcription factors which bind to specific sequence motifs and give prominent peaks. In such cases, the recommended sequencing depth is ~10–15 million reads, although the number of putative target regions identified tends to increase with the number of reads, especially for weaker peaks [49, 50]. In contrast, broad peaks usually require more sequencing depth (~30–50 million reads) [27, 51]. Modified histones, RNA polymerases, chromatin remodelers, and other factors display this type of distribution [52]. When studying a factor the distribution of which is unknown, a more quantitative approach is to empirically evaluate the saturation point (number of reads after which more sequencing does not help discovering more binding sites) [53]. Appropriate sequencing depth also depends greatly on the S/N ratio of the antibody.

5.4 Library Complexity

Peak detection and reproducibility of binding site discovery benefit greatly from a more complex library. Complexity of the library is measured by calculating the amount of reads that map to unique locations of the genome versus the total number of reads sequenced. This is known as the nonredundant fraction (NRF). With more sequencing depth, the nonredundant fraction of a library is eventually exhausted and the increase in reads only contributes to sequencing the same PCR-amplified fragments repeatedly [30, 54]. Low library complexities may result from low-input DNA or other library construction problems.

5.5 Quality Control and Mappability

The initial steps after obtaining the raw sequence reads are quality control and mapping high quality reads to the reference genome. Quality control includes several main steps: (1) sequencing adaptor clipping, (2) detection of GC composition biases in the sequences which might indicate contamination of the samples with extraneous DNA, (3) trimming of low-quality portions of reads, or (4) filtering of low-quality reads entirely. Once filtered, it is not always possible to assign high-quality reads to unique locations of the genome due to their repetitive nature. The regions of the genome that contain such sequences (e.g., repeats and transposons) are regions of "low mappability." Mappability is also referred to as "uniqueness" of the sequence. Mappability often depends on the read length and software algorithm chosen to assign reads to their genomic locations [55, 56]. Once mapping is completed, but before peak calling, alignment files need to be visually inspected in genomes browsers such as IGV [57] or UCSC [58], to get a qualitative sense of the quality of the ChIP-seq data. A good ChIP dataset should contain overall stronger and more symmetrical, bell-shaped accumulations of signal (peaks) compared to control samples. Visual inspection is also useful to roughly determine the binding type of our target protein: narrow, broad, or mixed. Even though visual inspection of specific loci is a good indication of quality, more quantitative genome-wide measurements of quality control are recommended. One of the most informative genome-wide quality metrics is the calculation of the fraction of mapped reads that are contained in peaks (FRiP) [59]. In general, the higher the fraction the better the ChIP. A FRiP of 1% is considered good for the ENCODE project data [30].

5.6 Peak Calling

The most commonly used approach to discover binding sites of target proteins in the genome is peak calling [60]. A peak in ChIP-seq is an accumulation of signal observed in the IP'ed sample that is significantly higher than the background signal at the same locus in a control sample (e.g., input DNA or IgG ChIP). Peak calling is a fast growing and evolving field, with plenty of options to choose from depending on the ChIP application, and more tools are available for narrow binding factors than for broad or mixed factors. Peak callers are designed to count the number of read tags on windows of a given size across the genome, and to evaluate whether such counts are significantly higher than the background. Peak lists are reported with statistical confidence estimates like probability values (p-values) or false discovery rates (FDR). There are dozens of peak callers, which make use of very different algorithms and statistical frameworks [61, 62]. Two of the most widely adopted packages are MACS [63] and SICER [64]. MACS (Model-based Analysis of ChIP- Seq data) is better suited for the identification of sharp peaks, whereas SICER (Spatial Clustering Approach for the

Identification of ChIP-enriched Regions) is more appropriate for finding broader peaks. For a detailed evaluation, comparison and benchmarking of several peak callers, *see* refs. 65–68.

5.7 Similarity Between Replicates

Once the genome has been scanned and enrichment peaks have been identified in all independent replicates of the experiment, it is recommended to evaluate how similar the peak lists obtained are by performing a Irreproducible Discovery Rate (IDR) test [69]. IDR assumes that those peaks that are reproducibly found in all replicates with high confidence values describe better the binding of the target protein than "noisier" peaks characterized by lower or different confidence values (e.g., high in one but low in the other, or low in all replicates). The IDR test takes lists of peaks, ranks them by *p*-value or *q*-value, and plots the confidence of the calls between pairs of replicates. In IDR plots, one can see how many peaks have been reliably called between pairs of replicates, thus getting a sense of the quality of the datasets. An IDR of 1% is usually set as threshold to discriminate between reproducible and irreproducible peak findings.

5.8 Downstream Applications

After obtaining reliable lists of enriched binding sites, there are many downstream analyses that can be performed to gain insights into the underlying biological implications of ChIP-seq data. The most common steps included in data analysis pipelines, as well as recommended software tools to perform them, are:

- Visual inspection of the peaks in genome browsers such as IGV.
- Plotting of gene-averaged binding distributions relative to genomic coordinates (e.g., promoters, enhancers, or transcription start sites) with ngs.plot [70] or SeqPlots [71].
- Clustering of individual binding patterns to find potentially co-regulated genes with the MeV suite [72].
- Differential peak analysis between time points or experimental conditions with DiffBind [73].
- Binding motif discovery with the MEME [74] or HOMER [54] suites.
- Finding genes and other features that overlap or neighbor our peaks with the PAVIS [75] or GREAT [76] suites.
- Gene ontology analysis of the genes bound by our target protein with GOrilla [77].
- Integration with other genomics datasets (e.g., RIP-seq, RNA-seq, GRO-seq, ChIP-chip) [78, 79].

5.9 Improving the Resolution of ChIP-Seq Data

Traditional ChIP-seq does not allow for single-nucleotide resolution mapping. This is due to the fact that the length of the DNA fragments immunoprecipitated by ChIP is determined by the method used to prepare chromatin (sonication or MNase digestion) and is usually much longer than the binding sequence of the protein of interest. With such relatively large fragments, it is not possible to reliably extract the DNA sequence contacted by the protein of interest. However, identifying the precise binding site can be of outmost importance, for example when considering that changes in the positions of nucleosomes of as little as 1 nt can have important implications for chromatin organization, or when the aim of the experiment is to discover the precise sequence motif bound by a transcription factor of interest. Several efforts have been made to improve the sequence resolutionChIP-exo of ChIP-seq [80]. One of them has resulted in the development of the ChIP-exo approach [29, 81, 82], in which sonicated and IP'ed crosslinked chromatin is adaptor-ligated and incubated with an enzyme (λ-exonuclease) that trims free DNA ends until it collides with a DNA-binding protein. In addition, λ-exonuclease fully digests all non-protein bound DNA fragments, improving the SN ratio of the method. An alternative method is High-Resolution ChIP [28], which combines MNase digestion with sonication. This strategy has the double advantage of helping solubilize proteins bound to certain genomic regions that would otherwise be underrepresented with only one method, and of increasing the resolution of the binding site discovery by making the DNA fragments shorter before IP.

6 Future Challenges: ChIP of Single Cells or Small Cell Numbers

ChIP data derive from averaging chromatin states from bulk population of cells. However, in recent years, a growing field of research has focused on studying the genomes and transcriptomes of smaller, more homogeneous populations, at single-cell resolution [83, 84]. This approach has revealed an unprecedented amount of cell-to-cell variation in homogeneous populations [85–87]. Single-cell omics methodologies based in DNA-seq and RNA-seq are instrumental for these discoveries, and are shaping our understanding of the chromatin landscapes from precious samples (e.g., primary cells or clinical samples). However, there is still no true single-cell ChIP methodology available, mainly due to the fact that chromatin profiling methods require high numbers of input material to avoid low S/N ratios and poor library complexity [88]. Standard requirements for ChIP-seq range from 10^6 to 10^7 cells. Such number of cells is feasible when working with cultured cells, with tissues or with complete organisms, but it might be more challenging to obtain in specific cases such as clinical samples. To overcome this limitation, continuous efforts are being made to increase the SN ratio of ChIP with low cell numbers. One approach takes advantage of the increased solubility of native chromatin preparation followed by MNase fragmentation to perform ChIP with as few as 10^4 cells [89]. When it comes to library preparation, a handful of methods have been devised that optimize the protocols for low input samples and reduce the need for PCR amplification [90, 91]. The most well-known are Nano-ChIP [92] and LinDA [93]. In parallel, advances in miniaturization and microfluidics have opened the possibility of reducing the cell numbers even more, from 10,000 cells [94, 95] to even single cells. However, the first reports have been very limited in terms of number of reads obtained per cell [96]. All these advances are taking the ChIP method to a new era and are opening new avenues for our understanding of the establishment, regulation, and biological significance of the interactions between DNA and proteins in chromatin.

Acknowledgments

Our work is supported by grants from The Swedish Research Council and The Swedish Cancer Society to N.V. A.J.P. was supported by the Department of Molecular Biosciences, The Wenner-Gren Institute at the Stockholm University.

References

1. van Steensel B (2011) Chromatin: constructing the big picture. EMBO J 30(10): 1885–1895. doi:10.1038/emboj.2011.135
2. MacAlpine DM, Almouzni G (2013) Chromatin and DNA replication. Cold Spring Harb Perspect Biol 5(8):1–22. doi:10.1101/cshperspect.a010207
3. Bannister AJ, Kouzarides T (2011) Regulation of chromatin by histone modifications. Cell Res 21(3):381–395. doi:10.1038/cr.2011.22
4. Zemach A, McDaniel IE, Silva P et al (2010) Genome-wide evolutionary analysis of eukaryotic DNA methylation. Science 328 (5980):916–919. doi:10.1126/science.1186366
5. Allis CD, Jenuwein T (2016) The molecular hallmarks of epigenetic control. Nat Rev Genet 17(8):487–500. doi:10.1038/nrg.2016.59
6. Gilmour DS, Lis JT (1984) Detecting protein-DNA interactions in vivo: distribution of RNA polymerase on specific bacterial genes. Proc Natl Acad Sci U S A 81:4275–4279. doi:10.1073/pnas.81.14.4275
7. Gilmour DS, Lis JT (1985) In vivo interactions of RNA polymerase II with genes of Drosophila melanogaster. Mol Cell Biol 5:2009–2018. doi:10.1128/mcb.5.8.2009
8. Solomon MJ, Larsen PL, Varshavsky A (1988) Mapping protein-DNA interactions in vivo with formaldehyde: evidence that histone H4 is retained on a highly transcribed gene. Cell 53 (6):937–947. doi:10.1016/S0092-8674(88)90469-2
9. Mayer A, Lidschreiber M, Siebert M et al (2010) Uniform transitions of the general RNA polymerase II transcription complex. Nat Struct Mol Biol 17(10):1272–1278
10. Kharchenko PV, Alekseyenko AA, Schwartz YB et al (2011) Comprehensive analysis of the chromatin landscape in Drosophila. Nature 471(7339):480–485. doi:10.1038/nature09725
11. Kasinathan S, Orsi GA, Zentner GE et al (2014) High-resolution mapping of transcription factor binding sites on native chromatin. Nat Methods 11(2):203–209. doi:10.1038/nmeth.2766
12. Kidder BL, Hu G, Zhao K (2011) ChIP-seq: technical considerations for obtaining high quality data. Nat Immunol 12(10):918–922. doi:10.1038/ni.2117
13. Rozowsky J, Euskirchen G, Auerbach RK et al (2009) PeakSeq enables systematic scoring of ChIP-seq experiments relative to controls. Nat Biotechnol 27:66–75

14. Leek JT, Scharpf RB, Bravo HC et al (2010) Tackling the widespread and critical impact of batch effects in high-throughput data. Nat Rev Genet 11(10):733–739. doi:10.1038/nrg2825
15. Hoffman EA, Frey BL, Smith LM et al (2015) Formaldehyde crosslinking: a tool for the study of chromatin complexes. J Biol Chem 290 (44):26404–26411. doi:10.1074/jbc.R115.651679
16. Darnell RB (2010) HITS-CLIP: panoramic views of protein-RNA regulation in living cells. Wiley Interdiscip Rev RNA 1 (2):266–286. doi:10.1002/wrna.31
17. Hafner M, Landthaler M, Burger L et al (2010) Transcriptome-wide identification of RNA-binding protein and microRNA target sites by PAR-CLIP. Cell 141(1):129–141. doi:10.1016/j.cell.2010.03.009
18. Solomon MJ, Varshavsky A (1985) Formaldehyde-mediated DNA-protein cross-linking: a probe for in vivo chromatin structures. Proc Natl Acad Sci U S A 82:6470–6474
19. Quievryn G, Zhitkovich A (2000) Loss of DNA-protein crosslinks from formaldehyde-exposed cells occurs through spontaneous hydrolysis and an active repair process linked to proteosome function. Carcinogenesis 21:1573–1580
20. Kennedy-Darling J, Smith LM (2014) Measuring the formaldehyde protein-dna cross-link reversal rate. Anal Chem 86(12):5678–5681. doi:10.1021/ac501354y
21. Turner B (2001) ChIP with native chromatin: advantages and problems relative to methods using cross-linked material. In: Mapping protein/DNA interactions by cross-linking [Internet]. Institut National de la Santé et de la Recherche Médicale, Paris
22. Gade P, Kalvakolanu DV (2012) Chromatin immunoprecipitation assay as a tool for analyzing transcription factor activity. Methods Mol Biol 809:85–104. doi:10.1007/978-1-61779-376-9_6
23. Mayer A, Churchman LS (2016) Genome-wide profiling of RNA polymerase transcription at nucleotide resolution in human cells with native elongating transcript sequencing. Nat Protoc 11(4):813–833. doi:10.1038/nprot.2016.047
24. O'Neill LP, Turner BM (2003) Immunoprecipitation of native chromatin: NChIP. Methods 31(1):76–82
25. Heins JN, Suriano JR, Taniuchi H et al (1967) Characterization of a nuclease produced by Staphylococcus aureus. J Biol Chem 242 (5):1016–1020
26. Dingwall C, Lomonossoff GP, Laskey RA (1981) High sequence specificity of micrococcal nuclease. Nucleic Acids Res 9:2659–2673
27. Chen Y, Negre N, Li Q et al (2012) Systematic evaluation of factors influencing ChIP-seq fidelity. Nat Methods 9:609–614
28. Skene PJ, Hernandez AE, Groudine M et al (2014) The nucleosomal barrier to promoter escape by RNA polymerase II is overcome by the chromatin remodeler Chd1. elife 3:e02042
29. Rhee HS, Pugh BF (2011) Comprehensive genome-wide protein–DNA interactions detected at single-nucleotide resolution. Cell 147:1408–1419
30. Landt SG, Marinov GK, Kundaje A et al (2012) ChIP-seq guidelines and practices of the ENCODE and modENCODE consortia. Genome Res 22(9):1813–1831. doi:10.1101/gr.136184.111
31. Peach SE, Rudomin EL, Udeshi ND et al (2012) Quantitative assessment of chromatin immunoprecipitation grade antibodies directed against histone modifications reveals patterns of co-occurring marks on histone protein molecules. Mol Cell Proteomics 11 (5):128–137. doi:10.1074/mcp.M111.015941
32. Park D, Lee Y, Bhupindersingh G et al (2013) Widespread misinterpretable ChIP-seq bias in yeast. PLoS One 8(12):e83506. doi:10.1371/journal.pone.0083506
33. Auerbach RK, Euskirchen G, Rozowsky J et al (2009) Mapping accessible chromatin regions using Sono-Seq. Proc Natl Acad Sci U S A 106 (35):14926–14931. doi:10.1073/pnas.0905443106
34. Jiang C, Pugh BF (2009) Nucleosome positioning and gene regulation: advances through genomics. Nat Rev Genet 10(3):161–172. doi:10.1038/nrg2522
35. Ehrensberger AH, Franchini DM, East P et al (2015) Retention of the native epigenome in purified mammalian chromatin. PLoS One 10 (8):e0133246. doi:10.1371/journal.pone.0133246
36. Bonhoure N, Bounova G, Bernasconi D et al (2014) Quantifying ChIP-seq data: a spiking method providing an internal reference for sample-to-sample normalization. Genome Res 24(7):1157–1168. doi:10.1101/gr.168260.113
37. Eberle AB, Böhm S, Östlund Farrants AK et al (2012) The use of a synthetic DNA-antibody complex as external reference for chromatin immunoprecipitation. Anal Biochem 426 (2):147–152. doi:10.1016/j.ab.2012.04.020

38. Buck MJ, Lieb JD (2004) ChIP-chip: considerations for the design, analysis, and application of genome-wide chromatin immunoprecipitation experiments. Genomics 83:349–360
39. He X, Cicek AE, Wang Y et al (2015) De novo ChIP-seq analysis. Genome Biol 16(1):205. doi:10.1186/s13059-015-0756-4
40. Huber W, Toedling J, Steinmetz LM (2006) Transcript mapping with high-density oligonucleotide tiling arrays. Bioinformatics 22 (16):1963–1970
41. Zhao S, Fung-Leung WP, Bittner A et al (2014) Comparison of RNA-Seq and microarray in transcriptome profiling of activated T cells. PLoS One 9(1):e78644
42. Furey TS (2012) ChIP-seq and beyond: new and improved methodologies to detect and characterize protein-DNA interactions. Nat Rev Genet 13:840–852. doi:10.1038/nrg3306
43. Goodwin S, McPherson JD, McCombie WR (2016) Coming of age: ten years of next-generation sequencing technologies. Nat Rev Genet 17(6):333–351. doi:10.1038/nrg.2016.49
44. Aird D, Ross MG, Chen WS et al (2011) Analyzing and minimizing PCR amplification bias in Illumina sequencing libraries. Genome Biol 12(2):R18. doi:10.1186/gb-2011-12-2-r18
45. Chabbert CD, Adjalley SH, Klaus B et al (2015) A high-throughput ChIP-Seq for large-scale chromatin studies. Mol Syst Biol 11(1):777. doi:10.15252/msb.20145776
46. van Galen P, Viny AD, Ram O et al (2016) A multiplexed system for quantitative comparisons of chromatin landscapes. Mol Cell 61 (1):170–180. doi:10.1016/j.molcel.2015.11.003
47. Weiner A, Lara-Astiaso D, Krupalnik V et al (2016) Co-ChIP enables genome-wide mapping of histone mark co-occurrence at single-molecule resolution. Nat Biotechnol 34 (9):953–961. doi:10.1038/nbt.3652
48. Meyer CA, Liu XS (2014) Identifying and mitigating bias in next-generation sequencing methods for chromatin biology. Nat Rev Genet 15:709–721
49. Myers RM, Stamatoyannopoulos J, Snyder M et al (2011) A user's guide to the encyclopedia of DNA elements (ENCODE). PLoS Biol 9: e1001046. doi:10.1371/journal.pbio.1001046
50. Sims D, Sudbery I, Ilott NE et al (2014) Sequencing depth and coverage: key considerations in genomic analyses. Nat Rev Genet 15:121–132
51. Jung YL, Luquette LJ, Ho JW et al (2014) Impact of sequencing depth in ChIP-seq experiments. Nucleic Acids Res 42:e74. doi:10.1093/nar/gku178
52. Bailey T, Krajewski P, Ladunga I et al (2013) Practical guidelines for the comprehensive analysis of ChIP-seq data. PLoS Comput Biol 9(11):e1003326. doi:10.1371/journal.pcbi.1003326
53. Kharchenko PV, Tolstorukov MY, Park PJ (2008) Design and analysis of ChIP-seq experiments for DNA-binding proteins. Nat Biotechnol 26(12):1351–1359. doi:10.1038/nbt.1508
54. Heinz S, Benner C, Spann N et al (2010) Simple combinations of lineage-determining transcription factors prime cis-regulatory elements required for macrophage and B cell identities. Mol Cell 38:576–589
55. Koehler R, Issac H, Cloonan N et al (2011) The uniqueome: a mappability resource for short-tag sequencing. Bioinformatics 27:272–274
56. Derrien T, Estelle J, Marco Sola S et al (2012) Fast computation and applications of genome mappability. PLoS One 7:e30377
57. Robinson JT, Thorvaldsdóttir H, Winckler W et al (2011) Integrative genomics viewer. Nat Biotechnol 29(1):24–26. doi:10.1038/nbt.1754
58. Tyner C, Barber GP, Casper J et al (2017) The UCSC genome browser database: 2017 update. Nucleic Acids Res 45(Dastabase issue):D626–D634. doi:10.1093/nar/gkw1134
59. Ji H, Jiang H, Ma W et al (2008) An integrated software system for analyzing ChIP-chip and ChIP-seq data. Nat Biotechnol 26:1293–1300
60. Diaz A, Park K, Lim DA et al (2012) Normalization, bias correction, and peak calling for ChIP-seq. Stat Appl Genet Mol Biol 11:article 9
61. Koohy H, Down TA, Spivakov M et al (2014) A comparison of peak callers used for DNase-Seq data. PLoS One 9(5):e96303. doi:10.1371/journal.pone.0096303
62. Thomas R, Thomas S, Holloway AK et al (2016) Features that define the best ChIP-seq peak calling algorithms. Brief Bioinform 18 (3):441–450. doi:10.1093/bib/bbw035
63. Zhang Y, Liu T, Meyer CA et al (2008) Model-based analysis of ChIP-Seq (MACS). Genome Biol 9(9):R137. doi:10.1186/gb-2008-9-9-r137
64. Zang C, Schones DE, Zeng C et al (2009) A clustering approach for identification of enriched domains from histone modification

ChIP-Seq data. Bioinformatics 25 (15):1952–1958
65. Laajala TD, Raghav S, Tuomela S et al (2009) A practical comparison of methods for detecting transcription factor binding sites in ChIP-seq experiments. BMC Genomics 10:618
66. Wilbanks EG, Facciotti MT (2010) Evaluation of algorithm performance in ChIP-seq peak detection. PLoS One 5:e11471
67. Malone BM, Tan F, Bridges SM et al (2011) Comparison of four ChIP-Seq analytical algorithms using rice endosperm H3K27 trimethylation profiling data. PLoS One 6:e25260
68. Rye MB, Saetrom P, Drablos F (2011) A manually curated ChIP-seq benchmark demonstrates room for improvement in current peak-finder programs. Nucleic Acids Res 39: e25
69. Li Q, Brown J, Huang H et al (2011) Measuring reproducibility of high-throughput experiments. Ann Appl Stat 5:1752–1779
70. Shen L, Shao N, Liu X et al (2014) ngs.plot: quick mining and visualization of next-generation sequencing data by integrating genomic databases. BMC Genomics 15:284. doi:10.1186/1471-2164-15-284
71. Stempor P, Ahringer J (2016) SeqPlots – interactive software for exploratory data analyses, pattern discovery and visualization in genomics. Wellcome Open Res 1:14. doi:10.12688/wellcomeopenres.10004.1
72. Howe EA, Sinha R, Schlauch D et al (2011) RNA-Seq analysis in MeV. Bioinformatics 27 (22):3209–3210. doi:10.1093/bioinformatics/btr490
73. Ross-Innes CS, Stark R, Teschendorff AE et al (2012) Differential oestrogen receptor binding is associated with clinical outcome in breast cancer. Nature 481:389–393
74. Bailey TL, Johnson J, Grant CE et al (2015) The MEME suite. Nucleic Acids Res 43(Web Server issue):W39–W49. doi:10.1093/nar/gkv416
75. Huang W, Loganantharaj R, Schroeder B et al (2013) PAVIS: a tool for peak annotation and visualization. Bioinformatics 29 (23):3097–3099. doi:10.1093/bioinformatics/btt520
76. McLean CY, Bristor D, Hiller M et al (2010) GREAT improves functional interpretation of cis-regulatory regions. Nat Biotechnol 28 (5):495–501. doi:10.1038/nbt.1630
77. Eden E, Navon R, Steinfeld I et al (2009) GOrilla: a tool for discovery and visualization of enriched GO terms in ranked gene lists. BMC Bioinformatics 10:48. doi:10.1186/1471-2105-10-48
78. Angelini C, Costa V (2014) Understanding gene regulatory mechanisms by integrating ChIP-seq and RNA-seq data: statistical solutions to biological problems. Front Cell Dev Biol 2:51. doi:10.3389/fcell.2014.00051
79. Eberle AB, Jordán-Pla A, Gañez-Zapater A et al (2015) An Interaction between RRP6 and SU(VAR)3-9 Targets RRP6 to Heterochromatin and Contributes to Heterochromatin Maintenance in Drosophila melanogaster. PLoS Genet 11(9):3–9. doi:10.1371/journal.pgen.1005523
80. Zentner GE, Henikoff S (2014) High-resolution digital profiling of the epigenome. Nat Rev Genet 15(12):814–827
81. Rhee HS, Pugh BF (2012) Genome-wide structure and organization of eukaryotic pre-initiation complexes. Nature 483 (7389):295–301. doi:10.1038/nature10799
82. Serandour AA, Brown GD, Cohen JD et al (2013) Development of an Illumina-based ChIP-exonuclease method provides insight into FoxA1-DNA binding properties. Genome Biol 14:R147
83. Grün D, van Oudenaarden A (2015) Design and analysis of single-cell sequencing experiments. Cell 163(4):799–810. doi:10.1016/j.cell.2015.10.039
84. Junker JP, van Oudenaarden A (2014) Every cell is special: genome-wide studies add a new dimension to single-cell biology. Cell 157 (1):8–11. doi:10.1016/j.cell.2014.02.010
85. Shalek AK, Satija R, Adiconis X et al (2013) Single-cell transcriptomics reveals bimodality in expression and splicing in immune cells. Nature 498(7453):236–240. doi:10.1038/nature12172
86. Kalisky T, Quake SR (2011) Single-cell genomics. Nat Methods 8:311–314
87. Munsky B, Neuert G, van Oudenaarden A (2012) Using gene expression noise to understand gene regulation. Science 336:183–187
88. Barski A, Cuddapah S, Cui K et al (2007) High-resolution profiling of histone methylations in the human genome. Cell 129 (4):823–837
89. Gilfillan GD, Hughes T, Sheng Y et al (2012) Limitations and possibilities of low cell number ChIP-seq. BMC Genomics 13:645. doi:10.1186/1471-2164-13-645
90. Schmidl C (2015) ChIPmentation: fast, robust, low-input ChIP-Seq for histones and transcription factors. Nat Methods 12:963
91. Bolduc N (2016) Preparation of low-input and ligation-free libraries using template-switching technology. In: Current protocols in molecular biology, vol unit 7.26. Wiley & Sons

92. Adli M, Bernstein BE (2011) Whole-genome chromatin profiling from limited numbers of cells using nano-ChIP-seq. Nat Protoc 6(10): 1656–1668. doi:10.1038/nprot.2011.402
93. Shankaranarayanan P, Mendoza-Parra MA, Walia M et al (2011) Single-tube linear DNA amplification (LinDA) for robust ChIP-seq. Nat Methods 8(7):565–567. doi:10.1038/nmeth.1626
94. Acevedo LG, Iniguez AL, Holster HL et al (2007) Genome-scale ChIP-chip analysis using 10,000 human cells. BioTechniques 43:791–797
95. Dahl JA, Collas P (2009) MicroChIP: chromatin immunoprecipitation for small cell numbers. Methods Mol Biol 567:59–74. doi:10.1007/978-1-60327-414-2_4
96. Rotem A, Ram O, Shoresh N et al (2015) Single-cell ChIP-seq reveals cell subpopulations defined by chromatin state. Nat Biotechnol 33(11):1165–1172

Chapter 3

How to Combine ChIP with qPCR

Patrik Asp

Abstract

Chromatin immunoprecipitation (ChIP) coupled with quantitative PCR (qPCR) has in the last 15 years become a basic mainstream tool in genomic research. Numerous commercially available ChIP kits, qPCR kits, and real-time PCR systems allow for quick and easy analysis of virtually anything chromatin-related as long as there is an available antibody. However, the highly accurate quantitative dimension added by using qPCR to analyze ChIP samples significantly raises the bar in terms of experimental accuracy, appropriate controls, data analysis, and data presentation. This chapter will address these potential pitfalls by providing protocols and procedures that address the difficulties inherent in ChIP-qPCR assays.

Key words ChIP, qPCR, Chromatin, Quantitative PCR, Primer design, Amplicon, qPCR calculations, ChIP-qPCR data presentation

1 Introduction

Simply put, a ChIP experiment compares the location-specific genomic DNA enrichment found in a target-specific chromatin immunoprecipitation with the enrichment found in an unspecific control sample. In the early days, this enrichment comparison was done by running location-specific PCRs (sometimes with radioactive nucleotides) on ChIP-enriched DNA and controls and compare the results on SDS-PAGE or agarose gels in a semi-quantitative manner [1]. This approach was labor-intensive and required manual determination of the linear range of the PCR reactions, making ChIP assays semi-quantitative at best. When qPCR or real-time PCR became a mainstream technology, it was quickly adopted as a means of accurately measuring ChIP enrichment in a truly quantitative manner compared to previous gel analyses [2–4].

With a few exceptions, chromatin immunoprecipitation is a low-efficiency assay that only captures a fraction of the possible targets present across the genome due to variability in cross-linking efficiencies and/or steric hindrance of antibody accessibility. Cross-

Neus Visa and Antonio Jordán-Pla (eds.), *Chromatin Immunoprecipitation: Methods and Protocols*,
Methods in Molecular Biology, vol. 1689, DOI 10.1007/978-1-4939-7380-4_3,

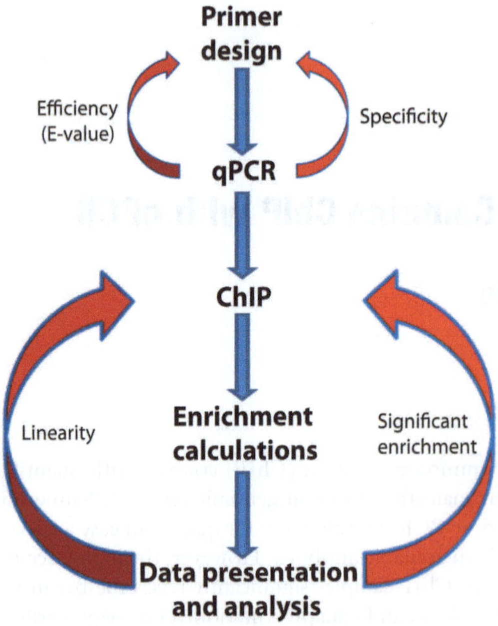

Fig. 1 A flow-chart of the many steps required to generate high-quality ChIP-qPCR data. *Blue arrows* indicate the standard linear stepwise progression through a ChIP-qPCR experiment while the *red arrows* highlight the various quality control steps and feedback loops required to ensure high-quality data

linked chromatin is also a highly complex biochemical substrate in terms of structure and molecular surface charges, and is therefore prone to high background signals as it "sticks" to beads, tubes, and pipette tips. Unfortunately, the ease by which ChIP-qPCR assays can now be performed creates an illusion of simplicity that masks the multitude of technical difficulties inherent with this assay.

To address the many parameters influencing ChIP-qPCR experiments, this chapter will follow the linear flow outlined in Fig. 1, discussing the five major steps involved in ChIP-qPCR (blue arrows) in parallel with the necessary quality control and feedback procedures associated with each of them (red arrows). Together, these detailed descriptions, discussions, and protocols will provide a clear roadmap toward obtaining high-quality ChIP-qPCR data.

2 Materials

Use molecular grade, purified water for all applications and sterile filter all solutions and buffers. Dissolve DNA (input, ChIP enriched DNA, primers, etc.) in 10 mM Tris pH 7.8–8 without EDTA as a slightly basic pH will limit chemical degradation of DNA. Excluding the EDTA chelator is important as even minute traces may negatively affect sensitive qPCR reaction by sequestering divalent cations necessary for full polymerase activity (*see* **Note 1**).

There are many different real-time PCR machines and systems available today and although they may work in slightly different ways and/or require different reagents, the underlying principles and the acquired Ct data (*C*ycle *t*hreshold) are the same. There are also many commercially available, high-quality qPCR reagents and kits available so choose the one recommended for the real-time PCR machine you plan to use and follow the manufacturer's instructions. The choice between basic SYBR green-based systems or more advanced and expensive probe-based systems depends on the technical specifics of the ChIP and quality controls and optimizations will be the same for both. As a "rule of thumb": ready-made qPCR master mixes make it easier to minimize variation and get consistent results between replicate samples.

Although good bench-work skills and practices are critical in minimizing the background from unspecific binding, it is essential to include unspecific antibody ChIP control reactions (IgG controls). IgG controls should ideally be pre-immune sera which unfortunately can be difficult to obtain for commercial antibodies. However, generic baseline IgG from non-immunized animals works well as negative control and is commercially available. Use the same total amount of antibody in the IgG control as in the target-specific ChIP reactions.

3 Methods

3.1 General Considerations

1. The software for many real-time PCR programs contains various forms of automatic built-in normalization procedures that could seriously skew your data unless care is taken. Make sure, therefore, that you understand how the software of your specific system determines the Ct values from the original real-time data.
2. Make sure the real-time PCR program runs to completion by setting the total cycle number to 45. This ensures that the PCR reaction spans the entire kinetic range, minimizing the risk of missing data from low abundance templates: very clean IP control samples may require higher cycle numbers before reaching the linear range and the final plateau phase.
3. If the program/machine allows it, it is good to always have the melt-curve module (*see* below) attached so that it runs after the qPCR has finished. Although adding a little bit of time, it is a simple quality control step. Primers that work well under normal conditions may sometimes give rise to unspecific amplicons/primer-dimers under low template level conditions (i.e., control ChIPs) that require high cycle numbers before producing detectable amplicon levels.

Table 1
Quantity and volume guidelines for ChIP-qPCR assays

Type of ChIP	Total chromatin (μg)	% qPCR input	Ab (μg)	Final sample volume (μl)
Low abundance TFs or indirect chromatin binders such as HDACs and similar	50	0.5	5–10	25
Regular TFs, RNA pol I and II, and similar	25	0.5	5	50
Histones, epigenetic histone marks, RNA pol III, and similar	3–10	0.5	2–5	50

The abundance of chromatin targets as well as their ability to be efficiently crosslinked determines how much chromatin and antibody is needed to produce a quantitative qPCR signal. The final volume in which the purified DNA is dissolved will of course also greatly impact signal strength and detection levels. Low or high qPCR template amounts lead to high or low Ct values that are less quantitative and may also have lower specificity. Input samples are taken from pre-cleared chromatin samples intended for control ChIPs and stored at −20 °C until processed in parallel with the ChIP-enriched samples. The "Final sample volume" assumes using 2 μl as template in a qPCR. *TF* transcription factor, *HDAC* histone deacetylase

4. Prepare master mixes when setting up qPCR reactions to minimize technical variability. Make master mixes for all samples containing the same primers, distribute into the PCR tubes/wells, and only pipette the individual template samples into each reaction.
5. Always run technical replicates of each qPCR by setting up three identical reactions for each sample.
6. Always perform and analyze every ChIP in three independent biological replicates and present the data with associated standard deviations to enable objective evaluation of the underlying biology and data quality (*see* **Note 2**).
7. Adjust the starting amounts of chromatin, antibody, and final sample volumes in order to have enough purified input and ChIP-enriched DNA to obtain high-quality Ct values. Low template concentrations due to large dissolving volumes will lead to high Ct values and increased technical variability (*see* **Note 3**). The numbers given in Table 1 are useful starting points but may of course have to be adjusted according to your specific needs.

3.2 Primer Design

In general, primer design for ChIP-qPCR is the same as for any qPCR application but with a few extra control steps added.

1. Design primers for both positive and negative ChIP targets in order to accurately evaluate the specificity of the ChIP assay.
2. Make sure the correct genomic region is targeted: promoters, coding regions, enhancers, etc., depending on the biology of what you are studying.

3. Design amplicons to be between 80 and 150 bp long as this gives the best kinetic linearity in a PCR and will improve quantitative amplification. There are many primer design programs freely available on-line such as Primer3 Web that all work well (*see* **Note 4**).
4. Test the specificity of the suggested primers by in silico PCR using the relevant genome as template. This is to ensure, as much as possible, that the primers only amplify the intended target. The UCSC Genome Browser has a simple in silico PCR function with a clear tabulated and graphic output format (*see* **Note 5**). Test the suggested primer pairs and only keep the ones that pass the in silico quality control.
5. Primer specificity is equally important when a probe-based qPCR approach is the assay of choice. Probe hybridization makes the signal detection highly specific but if the primers give rise to multiple amplicons, the quantitative aspect of the PCR reaction itself will suffer greatly, counteracting the purpose of using qPCR in the first place.

3.3 Primer and Amplicon Testing

Despite the precautions described in the primer design step, all primer pairs need to be empirically tested in an actual qPCR as they may behave differently in the real world. It is a good practice to do this even when using primers from published papers to ensure they work as intended under the conditions present in your lab. There are two principle properties that need to be determined for each primer pair before proceeding with qPCR: efficiency and specificity. Most real-time PCR machines and software allow for simultaneous determination of these parameters.

3.3.1 Primer Efficiency—Determination of E-Value

Efficiency or E-value is a property specific to each primer pair and is a measurement of how well the primers perform in a PCR. In short, an efficiency of 100% means that a twofold dilution of the template results in a difference of Ct (ΔCt) of 1. Differences in E-values can have a significant impact on the quantitative fold-change calculations due to the exponential nature of PCR amplification (*see* qPCR calculations below).

1. Set the melt-curve program module to run immediately after the qPCR program in order to test efficiency and specificity in the same qPCR reaction.
2. Set up qPCR reactions using a twofold dilution series of five samples of the purified and pre-cleared input chromatin. Make the dilution series so that middle sample represents the amount of input chromatin normally used in an experiment as shown in Table 1, most often 0.5% (*see* **Note 6**).
3. Set up three technical PCR replicates of the dilution series for each primer pair and run the qPCR program to completion

Table 2
Examples of typical E-values obtained by primers designed using Primer 3 Web

Gene name	Fwd	Rev	E-value
Klf16_upstream	CGCGGAAGTAGGA GACCAG	ACAGACTGCTGTCG GTTGG	2 (100%)
Nr4a1_upstream	GAGGGGAGGAGATC CTGTTC	TCCAGACCCACTAGG CTCAC	1.936 (93.6%)
Nup155_downstream	ATGATGGTGGCCTCT ACGTC	CTCAGAAAGGTCCGGG TACA	1.927 (92.7%)

(45 cycles) using the same conditions (annealing temp, PCR mix, etc.) as for real ChIP samples. Make sure that the "melt curve" module is set to run immediately after the regular qPCR program is finished. If that is not technically possible due to the specific real-time PCR system used, run a separate but identical experiment to obtain the melt-curves.

4. In parallel, set up three replicate qPCR reactions without any template for each primer pair in order to test for primer–dimer formation (*see* Subheading 3.3.2 below).
5. Use the software to determine the E-values for the three replicate dilution series and calculate the average E-value for every primer pair.
6. Repeat for all primers intended for real-time PCR use and make a database with the empirically determined average E-values. Discard primer pairs with E-values lower than 1.8 (80% efficiency) or higher than 2 (efficiency >100%; rare) and re-design and test new primer pairs for these targets. Table 2 shows examples of real E-values for three primers pairs designed for mouse ChIP-qPCR.

3.3.2 Primer Specificity: Melt/Dissociation-Curve Analysis

For qPCR to be quantitative primer pairs must only produce one single amplicon and the level of primer-dimerization has to be very low or undetectable. If multiple PCR products are being generated or the primers self-hybridize, the reaction is no longer quantitative for a specific amplicon even though the PCR reaction as a whole will still have a linear amplification phase and measurable Ct value. This is why fluorescent probe-based assays also require highly specific primer pairs and low primer–dimer formation even though signal detection is specific for a unique amplicon. Therefore, a melt-curve or dissociation-curve analysis has to be performed for all primer pairs intended for use in a qPCR (*see* **Note** 7).

1. Run the melt/dissociation-curve analysis on the triplicate PCR reactions as described in the previous section. Only one ΔRFU

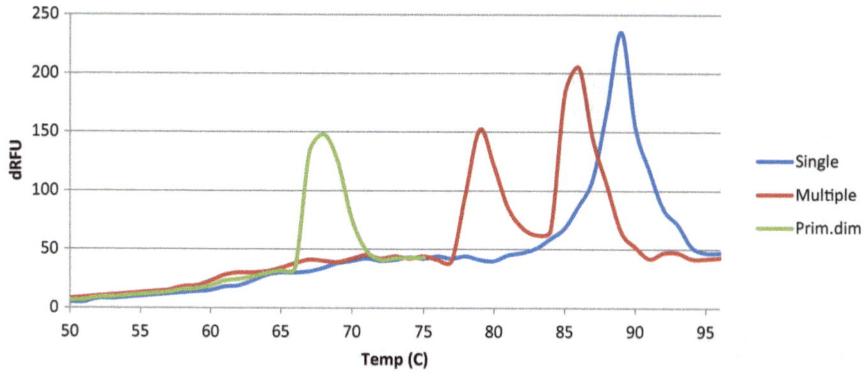

Fig. 2 Melt-dissociation-curve example. The blue reaction has one ΔRFU peak showing that the primer pair is specific and produces one PCR product. The red sample, however, displays two distinct ΔRFU peaks indicative of multiple amplicons and this primer pair would be discarded. Primer–dimers have a ΔRFU peak at slightly lower temperature due to their shorter lengths (*green sample*). In general, about 25–33% of all primer pairs will fail the E-value or melt-curve test (or both) and will have to be re-designed. Although it may seem tedious, quality testing of qPCR primers is a simple way of ensure accuracy and quality and to maximize confidence in downstream data analysis and interpretation

peak (relative fluorescence unit) will be present if the primer pair is specific (Fig. 2, blue sample) while two or more peaks will be visible if the primers are unspecific (Fig. 2, red sample). Discard unspecific primer pairs and re-design and test a new pair for that target.

2. Primer–dimer formation is detected in melt-curves form PCR reactions without template DNA, only primer DNA. Because double-stranded primer–dimer structures are shorter than PCR products they will have a lower melting temperature. Primer–dimer-generated ΔRFU peak will therefore show up at lower temperatures as shown in Fig. 2 (compare the blue and red samples with the green sample). Discard primer pairs that show dimer formation and re-design and test a new pair for that target (*see* **Note 8**).
3. For and extra layer of quality control, run the PCR reactions on a 2% agarose gel to make sure that the amplicon you see in the melt-curve analysis is of the correct size. Although unlikely, it is possible for a primer pair to produce a single amplicon that is of the wrong size and therefore incorrect. As before, discard such primer pairs and re-design and test a new pair for that target.

3.4 qPCR Calculations

Data output from real-time PCR software is in the form of Ct values (Cycle threshold): the cycle number at which point the reaction is in the linear range of the exponential PCR amplification process. It is the only time when the increase in PCR product in each cycle is directly proportional to the amount of starting material in the reaction. Differences in Ct values (ΔCt) between samples in the

A.

Sample	Ct	ΔCt to input	Fold difference (E=1.875)	% input
MB 0.5% input	30.36	0	1.00	0.50
MB IgG	34.86	-4.5	0.06	0.03
MB Pol II	31.67	-1.31	0.44	0.22
MT 0.5% IN	30.67	0	1.00	0.50
MT IgG	36.16	-5.49	0.03	0.02
MT Pol II	36.63	-5.96	0.02	0.01

B.

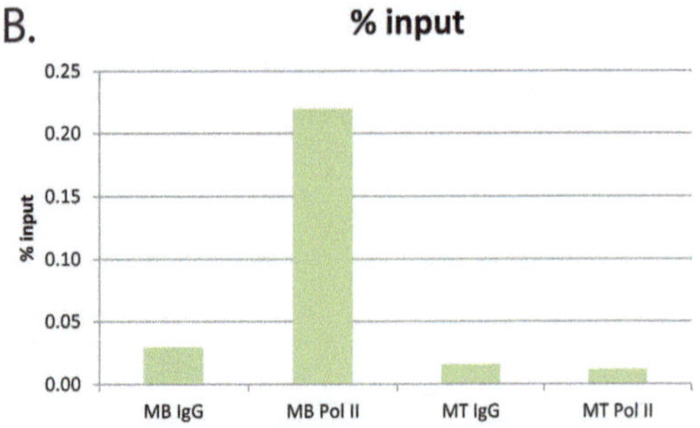

Fig. 3 Example of ChIP-qPCR calculations and data presentation. (**a**) ChIP-qPCR data analyzing Pol II occupancy on the mouse Top2a promoter before and after myogenic differentiation from myoblasts (MB) to myotubes (MT). (**b**) The result from the calculations in (**a**) plotted as % input. The data demonstrate that Pol II is absent from the Top2a promoter in MT, consistent with the fact that the Top2a gene is turned off during the differentiation process. The ChIP was titrated to be in the linear range and the significant enrichment cut-off is 0.05%

linear phase can therefore be used to calculate the relative fold-difference in the starting amount of DNA between samples. In the case of ChIP, the comparison is made between input, control, and ChIP samples as demonstrated by the ChIP-qPCR example shown in Fig. 3.

1. Calculate the ΔCt value between the input and the test samples: subtract the sample Ct value from the input Ct value (Fig. 3a). The input is the only sample with a known quantity of template DNA to which all other unknown samples will be compared.
2. Calculate the fold difference that the ΔCt values represent using the formula: $f = E^{\Delta \mathrm{Ct}}$, where E is the empirically determined efficiency value specific for each primer pair (Fig. 3a) [5].
3. Calculate the enrichment efficiency in each sample as the percentage of the starting amount of chromatin used for the ChIP by multiplying the fold-change number from **step 2** with the percentage used for the input sample (Fig. 2a).
4. Plot the % input values of the control IgG and sample ChIPs to visualize the result (Fig. 3b).

3.5 Determination of Significant Enrichment

The ChIP-qPCR data shown in Fig. 3 look convincing but it is important to recognize that it is only valid if the 0.22% enrichment seen in the MB sample is significant (*see* **Note 9**). ChIP assays are generally inefficient (with a few notable exceptions described below), prone to high background and exquisitely dependent on antibody specificity. Therefore, a significant enrichment threshold expressed as % input has to be empirically determined for each specific antibody/chromatin combination. This is done by a side-by-side comparison of enrichment signals from wild-type cells versus ChIP target negative control cells. The best control is a knock-out cell line deficient for the specific ChIP target while the second best choice is to use efficient RNA interference (siRNA) strategies to knock-down the target protein of the ChIP. A third option is to use a naturally occurring situation like the one described for the Top2a promoter above where the ChIP target is known to be present/absent depending on a well-known biological process.

1. Compare and verify the presence and absence of your target by Western Blot of wt and knock-out or knock-down cells/tissue. This is especially important when using siRNA-mediated approaches because target protein depletion has to be very high or it will not be possible to determine the significant enrichment threshold.
2. Run standard ChIP assays on chromatin from these cells/tissues and analyze with qPCR. Do not forget the IgG control ChIPs.
3. Calculate results as % input (*see* Subheading 3.4) and plot in a bar graph.
4. The residual signal you see in knock-out or knock-down samples should be close to the control IgG signal and represents the significant enrichment threshold. In general, this threshold is usually between 0.025 and 0.05% input but it can vary between antibodies. Keep in mind that the significance threshold is related to the total amount of input chromatin used and has to be re-titrated if the amount is increased.

3.6 Determining the Linearity of ChIP Assays

ChIP is a low-efficiency assay so in most cases saturation effects are not an issue. In fact, the opposite is often true and may require an increase in both chromatin and antibody amounts to achieve significant enrichment. The exception is when analyzing histones or histone marks such as H3K4me3 or similar which are highly abundant, stable, and efficiently cross-linked to DNA. In these cases, it is possible to saturate the ChIP reaction so that it is no longer linear or quantitative. This problem is mitigated by optimizing the chromatin and chromatin/antibody ratio by titrating the ChIP assay. This has to be done for every antibody/target combination where high efficiency is expected.

1. Lower the antibody amount to 1–2 μg/reaction. Start with 2 μg but be prepared to re-run the titration using lesser amounts.
2. Set up a series of ChIP reactions using increasing amounts of chromatin from 3 to 25 μg while keeping the antibody constant. Remember to set up identical IgG control reactions.
3. Analyze the result on a qPCR and plot the enrichment as % input as shown in Fig. 4.
4. Analyze multiple targets and determine the chromatin amount that is in the middle of the linear range which most likely will cover all targets tested and use this specific chromatin/antibody

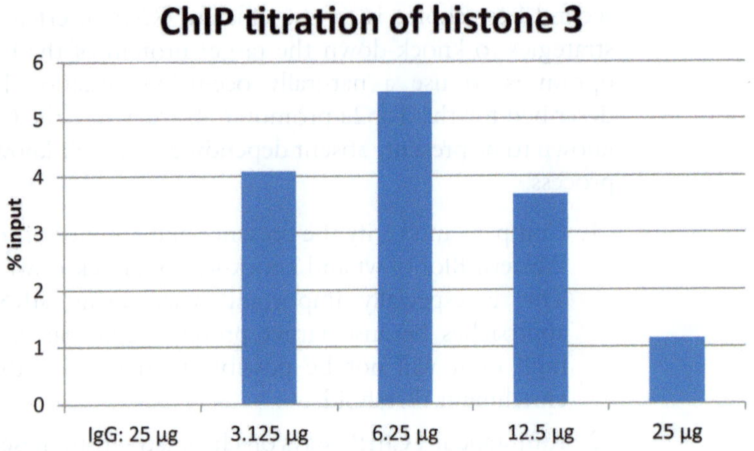

Fig. 4 ChIP titration of histone 3 occupancy on the mouse Notch 2 promoter in myoblast cells. 1 µg of antibody was used with increasing amounts of chromatin and the result showed that 3 µg of chromatin was optimal and therefore used in all subsequent H3 ChIPs. Note that the IgG signal shown is from the highest 25 µg chromatin ChIP in order to show the strongest possible background signal for a fair comparison

combination for all future experiments with this antibody. In the case of the data shown in Fig. 4, we used 1 µg of antibody and titrated increasing amounts of chromatin. This revealed a signal maximum with 6.25 µg chromatin but loss of signal at higher chromatin amounts. For this specific Ab/chromatin combination, we therefore decided to use 3.125 µg of chromatin and 1 µg of antibody for our experimental work. This combination yielded robust enrichment but was far enough from the conditions that gave rise to signal maximum and subsequent inhibiting saturation effects that the dynamic range of the ChIP assay would be able to reflect true biological changes free from technical artifacts.

3.7 ChIP-qPCR Data Presentation

All the tests and controls described above are pointless unless the ChIP-qPCR data are presented correctly. Although there are many differing opinions on how to present such data, one fundamental scientific principle always applies and may not be compromised: it must be possible to objectively evaluate the quality and validity of the quantitative data analysis (*see* **Note 10**).

1. Significant ChIP-qPCR enrichment cut-off values and linear range optimization ratios must be presented in some way. Either given in Materials & Methods, Supplemental data or verbally during an oral presentation.
2. Present IgG control data side-by-side with target specific data. It is common to see ChIP-qPCR results presented as the ratio of sample/IgG but that makes it impossible to objectively evaluate the quality of a ChIP by “hiding” the original IgG and sample

A.

	Sample	Ct	ΔCt to input	Fold difference (E=1.875)	% input	Fold over IgG
	0.5% input	30.36	0	1.00	0.50	N/A
	<u>IgG-2</u>	<u>32.86</u>	<u>-2.5</u>	<u>0.21</u>	<u>0.10</u>	<u>2.11</u>
	IgG-1	33.86	-3.5	0.11	0.06	3.96
MB	IgG	34.86	-4.5	0.06	0.03	7.43
	IgG+1	35.86	-5.5	0.03	0.02	13.93
	IgG+2	36.86	-6.5	0.02	0.01	26.11
	Pol II	31.67	-1.31	0.44	0.22	N/A
	0.5% IN	30.67	0	1.00	0.50	31.53
	IgG-2	34.16	-3.49	0.11	0.06	0.21
	IgG-1	35.16	-4.49	0.06	0.03	0.40
MT	IgG	36.16	-5.49	0.03	0.02	0.74
	IgG+1	37.16	-6.49	0.02	0.01	1.40
	<u>IgG+2</u>	<u>38.16</u>	<u>-7.49</u>	<u>0.01</u>	<u>0.00</u>	<u>2.62</u>
	Pol II	36.63	-5.96	0.02	0.01	N/A

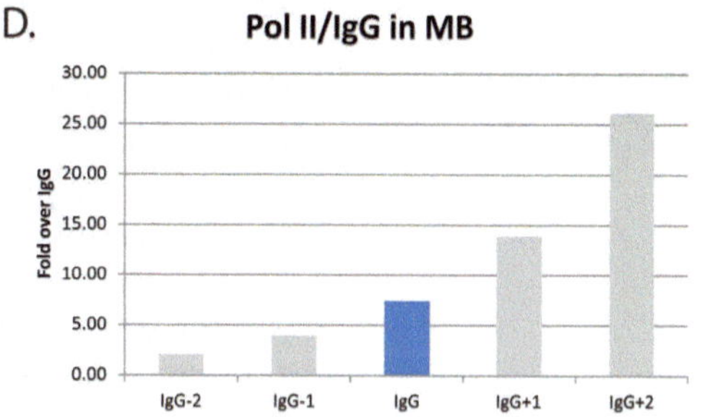

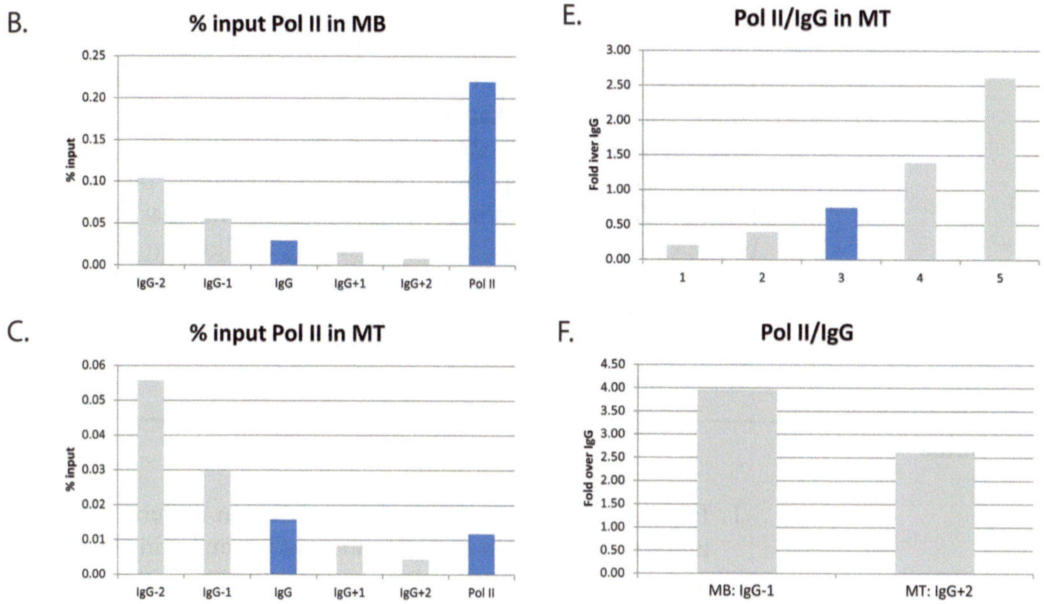

Fig. 5 ChIP-qPCR data presentation: separate IgG versus sample/IgG ratios. (**a**) Simulated IgG background values (*grey*) 1 and 2 Ct:s above and below the original IgG respectively were added to the data from Fig. 3. (**b**, **c**) MB and MT data as % input with original (*blue bars*) and virtual (*grey bars*) IgG data presented side-by-side with original sample data. (**d**, **e**) The same data as in (**b**, **c**) but presented as the ratio of sample/IgG. (**f**) MB and MT Pol II/IgG ratios using the highest and lowest simulated MB and MT IgG Ct values respectively

values and may also introduce significant bias in the data interpretation (Fig. 5).

The importance of correct data presentation is illustrated in Fig. 5. For educational purposes, simulated IgG background values 1 and 2 Ct:s above and below the original IgG were added to the original data shown in Fig. 3 (Fig. 5a, grey boxes). Note that in the legend for Fig. 3 it states that the Pol II ChIP had been titrated to be in the linear range and that the significance cut-off is 0.05%.

Presenting ChIP-qPCR data as % input (Fig. 5b, c) together with specifying the significance threshold allows for an objective evaluation of the data which is not possible if the data are presented as the sample over IgG ratio. Imagine for instance that the original

IgG result in MB cells looked like the simulated IgG-1 or IgG-2 samples in Fig. 5b. Both of these "controls" are above the significance threshold, indicative of high background in the experiment which puts the overall result into question. Importantly, if the same data are presented as sample over IgG as in panel 5d, it is no longer possible to independently evaluate the quality of the data because the actual IgG enrichment is no longer visible. All you can really say is that the sample enrichment must be higher than the IgG since the ratio is above 1. The real problem, however, occurs when using sample over IgG ratios to compare Pol II occupancy on the Top2a promoter. Imagine for example that the "real" experimental outcome was like that presented in Fig. 5f using the simulated underlined IgG values from the table in panel 5a. The conclusion would be that there is only a minor difference in Pol II occupancy on the Top2a promoter before and after myogenic differentiation, the opposite of what is known to be true. The fourfold difference in the amount of starting material represented by the Ct difference of 2 between the original IgG samples and the simulated samples shown in 5f is not unusual in any way. This clearly illustrates why it is critical to present IgG and sample enrichments individually instead of as fold ratios.

4 Notes

1. EDTA used to be a necessary component in buffers in order to prevent degradation of DNA due to contaminating enzymatic activities. However, this is no longer a problem with the molecular biology grade chemicals and reagents available today.
2. Unfortunately, the terms "technical" and "biological" replicates are often misunderstood: A technical replicate in a ChIP-qPCR is to run the same ChIP sample in three different but otherwise identical qPCR reactions. A biological replicate is when identical ChIPs are performed with chromatin from different replicate experiments. Producing consistent ChIP results across biological replicates is challenging and requires a lot of optimization and practice to achieve so running a total of nine qPCR reactions for each ChIP (3 biological replicates $\times$ 3 technical replicates) will significantly help in obtaining robust data with high statistical significance.
3. As a rule of thumb, the amount of template DNA in an input sample (0.25–1% of the total input chromatin) should give a Ct value of approximately 26–31 ($\pm$ a couple of cycles). Dissolve all other samples in the same volume as the input sample in order to directly compare them in a qPCR.
4. Primer 3 Web: http://bioinfo.ut.ee/primer3/. Identify and paste the target sequence into the sequence window, limit the

amplicon size to 80–150 bp, set the annealing temperature to 60 and let the program select potential primer pairs.

5. UCSC Genome Browser: http://genome.ucsc.edu/. Go to *Tools* and In-silico PCR, set the correct species, copy-and-paste forward and reverse primers and click "submit." The program will output an amplicon table and if the primer pair is specific, there will only be on single amplicon in the list. The name for each amplicon is a hyperlink that will show a graphic representation of the result. This is very useful as it will show the genomic location of the amplicon and verify that the primer pair amplifies the correct target.
6. Assuming the standard input is 0.5%, set up a dilution series containing 2, 1, 0.5, 0.25, and 0.125% input, respectively, in the same final sample volume used for a real ChIP-qPCR as shown in Table 1.
7. Most (if not all) qPCR systems are equipped with some form of melt-curve analysis function that first cools the finished PCR reaction to approximately 4 °C to ensure complete hybridization of double-stranded PCR products and then gradually increases the temperature up to 96 °C while simultaneously measuring the fluorescence signal. A melt-curve analysis then plots the changes in the signal (ΔRFU) as a function of increasing temperature. SYBR green and other dyes or probes used in real-time PCR only fluoresce when bound to double-stranded DNA. When the temperature reaches the specific melting point of a double-stranded PCR product, the fluorescent signal will therefore decrease drastically. Because different double-stranded DNA segments have different thermodynamic stability properties, they will generate ΔRFU peaks at different temperatures so nonspecific primer pairs will therefore produce more than one such peak.
8. In some cases, it is possible to neutralize primer–dimer formation by lowering the amount of primer in the qPCR. However, this may alter the kinetics of the PCR so the efficiency and specificity assays will have to be re-run with such new PCR conditions.
9. In this specific case, the absence of a Pol II signal in the MT sample coupled with the known biology of the Top2a genes convincingly argues that the enrichment is significant but it is not that common to have such a "built-in" quality control situation in a ChIP experiment.
10. This does not refer to the technical factors affecting qPCR specificity and quality, such as primer design, amplicon testing as it has to be assumed that these verifications have been done correctly or the entire experiment is invalid.

References

1. Ren B, Cam H, Takahashi Y et al (2002) E2F integrates cell cycle progression with DNA repair, replication, and G(2)/M checkpoints. Genes Dev 16(2):245–256
2. Asp P, Acosta-Alvear D, Tsikitis M et al (2009) E2f3b plays an essential role in myogenic differentiation through isoform-specific gene regulation. Genes Dev 23(1):37–53
3. Asp P, Blum R, Vethantham V et al (2011) Genome-wide remodeling of the epigenetic landscape during myogenic differentiation. Proc Natl Acad Sci U S A 108(22):E149–E158
4. Eberle AB, Jordan-Pla A, Ganez-Zapater A et al (2015) An interaction between RRP6 and SU(VAR)3-9 targets RRP6 to heterochromatin and contributes to heterochromatin maintenance in Drosophila melanogaster. PLoS Genet 11(9):e1005523
5. Pfaffl MW (2001) A new mathematical model for relative quantification in real-time RT-PCR. Nucleic Acids Res 29(9):e45

Chapter 4

Analysis of Protein–DNA Interaction by Chromatin Immunoprecipitation and DNA Tiling Microarray (ChIP-on-chip)

Hui Gao and Chunyan Zhao

Abstract

Chromatin immunoprecipitation (ChIP) has become the most effective and widely used tool to study the interactions between specific proteins or modified forms of proteins and a genomic DNA region. Combined with genome-wide profiling technologies, such as microarray hybridization (ChIP-on-chip) or massively parallel sequencing (ChIP-seq), ChIP could provide a genome-wide mapping of in vivo protein–DNA interactions in various organisms. Here, we describe a protocol of ChIP-on-chip that uses tiling microarray to obtain a genome-wide profiling of ChIPed DNA.

Key words Chromatin immunoprecipitation, Microarray, ChIP-on-chip, Protein–DNA interaction, Genome-wide

1 Introduction

Chromatin immunoprecipitation (ChIP) has become the most effective and widely used tool to study the interactions between specific proteins or modified forms of proteins and a genomic DNA region. In the modern ChIP procedure, formaldehyde as a reversible cross-linking agent was first used to form the interaction complex in live cells by cross-linking protein–DNA. The cross-linked chromatin was then sheared into small fragments by sonication, followed by immunoprecipitation with the desired antibodies. After the reversal of crosslink, the protein bound DNA was purified and these DNA sequences were identified via various methods.

During the last two decades, ChIP followed by genome-wide profiling technologies, such as microarray hybridization (ChIP-on-chip) or massively parallel sequencing (ChIP-seq), has enabled the analysis of many important DNA-interacting proteins including RNA polymerases, transcription factors, transcriptional co-factors, and histone proteins [1]. These genome-wide ChIP analysis

Neus Visa and Antonio Jordán-Pla (eds.), *Chromatin Immunoprecipitation: Methods and Protocols*,
Methods in Molecular Biology, vol. 1689, DOI 10.1007/978-1-4939-7380-4_4, © Springer Science+Business Media LLC 2018

approaches have led to many important discoveries related to transcriptional regulation [2, 3], epigenetic regulation [4], and have greatly enhanced our understanding of the organization and function of the mammalian genome.

Before the advent of next-generation sequencing, ChIP-on-chip was the standard technique for performing genome-wide mapping of protein–DNA interactions in various organisms. Generally, the purified ChIPed DNA is amplified, labeled to produce fluorescent signal, and hybridized to DNA microarrays containing genomic DNA sequences. The tiling microarray, a subtype of microarray, has been extensively used to localize DNA–protein interactions after ChIP. Unlike the traditional design of the microarray in which the probes for targeting sequences of known or predicted genes may be dispersed throughout the genome, tiling microarray is composed of a large number of probes that are immediately adjacent to (or overlap) one another from a contiguous region of the genome [5]. The fluorescent signals from ChIP DNA are subsequently identified and mapped back to the genome to identify the DNA interaction sites of the protein under study [6]. However, for many organisms it is not practical to generate commercially available genome-wide tiling arrays, and hence ChIP-on-chip data sets were often not genome-wide. Furthermore, the ability to detect binding sites in a ChIP-on-chip experiment is limited by the resolution of the probes on the array. Finally, the signal obtained by hybridization intensities on an array is analogical, and it is often difficult to determine levels of enrichment that are statistically significant and hence indicative of true binding sites [4]. Many of these limitations can be overcome by using ChIP-seq, since direct sequencing of the ChIPed DNA is not limited in any way by probes and it generally produces profiles with higher spatial resolution, dynamic range, and genomic coverage. Most importantly, ChIP-seq can be used to analyze virtually any species with a sequenced genome since it is not constrained by the availability of an organism-specific microarray.

Many variations of ChIP protocols have been developed, tested, and can be found in numerous literatures and websites [7–11]. In particular, the guidelines for ChIP from ENCODE and modENCODE consortia [12] and the beginner's guide to ChIP presented at Abcam website [13] contain very detailed ChIP protocols and excellent tips for designing experiments and troubleshooting.

ChIP-seq analysis has largely overtaken ChIP-on-chip as a mainstream method in genomics and epigenomics. Many vendors have stopped their production of the microarrays for ChIP analysis. Here, we present a ChIP-on-chip protocol with still commercially available Affymetrix tiling arrays. The final steps of Affymetrix tiling array analysis, such as hybridization of labeled DNA and scanning of the array, are usually performed at a core facility following very detailed proprietary protocols published elsewhere [14], this chapter will focus on providing a detailed wet lab protocol of the ChIP procedure, amplification, and labeling of the ChIPed DNA. The bioinformatics analysis of ChIP-on-chip data is not covered by this protocol, but there are also dedicated literatures and protocols on this subject elsewhere [15].

2 Materials

2.1 Chromatin Immunoprecipitation

1. Antibody for the protein of interest (*see* **Note 1**).
2. Negative control IgG.
3. Formaldehyde (37%).
4. 2.5 M Glycine.
5. cOmplete™, EDTA-free, protease inhibitor cocktail (Sigma).
6. Protein G Sepharose and Protein A Sepharose.
7. Fish sperm DNA.
8. BSA.
9. Phosphate-buffered saline (PBS): 137 mM NaCl, 2.7 mM KCl, 10 mM Na_2HPO_4, 1.8 mM KH_2PO_4.
10. Lysis Buffer 1: 50 mM Hepes-KOH, pH 7.5, 140 mM NaCl, 1 mM EDTA, 10% glycerol, 0.5% Igepal CA-630, 0.25% Triton X-100.
11. Lysis Buffer 1 (with 1× protease inhibitor cocktail): freshly made by dissolving one tablet of cOmplete™, EDTA-free, protease inhibitor cocktails in 7 mL Lysis Buffer 1.
12. Lysis Buffer 2: 200 mM NaCl, 1 mM EDTA, 0.5 mM EGTA, 10 mM Tris, pH 7.5.
13. Lysis Buffer 2 (with 1× protease inhibitor cocktail): freshly made by dissolving one tablet of cOmplete™, EDTA-free, protease inhibitor cocktails in 7 mL Lysis Buffer 2.
14. Lysis Buffer 3: 1 mM EDTA, 0.5 mM EGTA, 10 mM Tris–HCl, pH 7.5, 100 mM NaCl, 0.1% Na-Deoxycholate, 0.5% N-lauroylsarcosine.
15. Lysis Buffer 3 (with 1× protease inhibitor cocktail): freshly made by dissolving one tablet of cOmplete™, EDTA-free, protease inhibitor cocktails in 7 mL Lysis Buffer 3.
16. TSE I: 20 mM Tris–HCl, pH 7.5, 1% Triton X-100, 150 mM NaCl, 0.1% SDS, 2 mM EDTA.
17. TSE II: 20 mM Tris–HCl, pH 7.5, 1% Triton X-100, 500 mM NaCl, 0.1% SDS, 2 mM EDTA.
18. LiCl buffer: 20 mM Tris–HCl, pH 7.5, 1 mM EDTA, 250 mM LiCl, 1% Igepal CA-630, 1% Sodium deoxycholate.

19. TE buffer: 10 mM Tris–HCl, pH 7.5, 1 mM EDTA.
20. Elution Buffer: 1% SDS in TE buffer.
21. PCR purification kit.
22. Sonicator.
23. Rotating mixer with an adapter for 1.5 mL microcentrifuge tubes.

2.2 DNA Amplification, Labeling

1. Sequenase Version 2.0 DNA polymerase (USB).
2. Primer A: 200 mM 5′-GTTTCCCAGTCACGGTC(N)9-3′ (HPLC Purified).
3. Primer B: 100 mM 5′-GTTTCCCAGTCACGGTC-3′ (HPLC Purified).
4. Taq Polymerase 5 U/mL.
5. 100 mM each of dATP, dCTP, dGTP, dTTP, and dUTP.
6. BSA (20 mg/mL).
7. 1 M DTT (dithiothreitol).
8. 2 U/μL Uracil-DNA Glycosylase (UDG) (USB).
9. Human Apurinic/Apyrimidinic Endonuclease 1 (APE 1) (Includes 10× APE 1 Reaction Buffer) (USB).
10. 30 U/μL Terminal Deoxynucleotidyl Transferase, Recombinant (rTDT) (USB).
11. DNA Labeling Reagent, DLR, 10 mM (USB).

3 Methods

3.1 Chromatin Immunoprecipitation

To start with a large amount of cells is a key step for a success in the ChIP assay. As described in this section, it is usually sufficient to use 10–20 million cells, roughly one 15 cm culture dish of adherent cells, for each IP reaction. For every IP reaction with the antibody of interest, a control with nonspecific antibody will be needed.

1. Aspirate culture medium and add 19 mL of freshly made PBS containing 1% formaldehyde to each plate. Incubate for 10 min at room temperature with gentle shaking on an orbital shaker (*see* **Note 2**).
2. Quench crosslinks by directly adding glycine to a final concentration of 0.125 M and incubate for 5 min in the orbital shaker.
3. Aspirate the formaldehyde-glycine PBS into a waste container and wash the cells twice with 20 mL of ice-cold PBS. Aspirate after the second wash to remove all the remaining PBS.

4. Add 0.6 mL ice-cold PBS to each plate and scrape cells with a cell scraper. Transfer the cells to a 1.5 mL microcentrifuge tube and centrifuge at 2500 × *g* for 3 min to pellet the cells.
5. After the spin, aspirate to remove the supernatant. The pellet can be frozen at −80 °C for later use or continue to next step.
6. Resuspend each tube of cells in 1 mL of Lysis Buffer 1. To obtain the fully suspended cells, pipet up and down to dispense the cell pellets. Place the tubes on a rotating mixer for 10 min at 4 °C. Then spin at 2500 × *g* in a table top centrifuge for 5 min at 4 °C.
7. Aspirate the supernatant and resuspend each tube of pellet in 1 mL of Lysis Buffer 2. To obtain the fully suspended cells, pipet up and down to dispense the cell pellets. Place the tubes on a rotating mixer for 5 min at 4 °C. Then spin at 2500 × *g* in a table top centrifuge, 5 min at 4 °C.
8. Aspirate the supernatant and resuspend each tube of pellet in 600 μL of Lysis Buffer 3. To obtain the fully suspended cells, pipet up and down to dispense the cell pellets. Place the tubes on a rotating mixer for 5 min at 4 °C.
9. Sonicate the samples using a Branson Sonifier 250 with the condition of 10 cycles of sonication using microtip at power 2.5 with 90% duty for 10 s. It is important to keep samples cold during sonication. The samples could be immersed in a salt/ice/water bath (*see* **Note 3**).
10. Add 1/10 of 10% Triton X-100 to the lysate (600 μL of sample, 60 μL of 10% Triton X-100 to Lysis Buffer 3). Spin at 18,000 × *g* in a tabletop centrifuge for 10 min at 4 °C to pellet debris. Collected supernatant can be snap frozen and stored at −80 °C or proceed directly to the immunoprecipitation step (*see* **Note 4**).
11. Remove 60 μL (10% v/v) of sonicated DNA as total input. A small aliquot of the input can also be used to determine the extent of DNA shearing by analyzing the sizes of DNA fragments with gel electrophoresis or Agilent Bioanalyzer (*see* **Note 3**).
12. Prepare 50% of Protein G Sepharose and Protein A Sepharose bead slurry for pre-cleaning and immunoprecipitation. For each sample, mix 40 μL of each bead type together into a 1.5 mL non-sticky microcentrifuge tube. Take out a 20% extra of each bead type to compensate the loss during pipetting of the beads and tube transfer. Pellet the beads by centrifuging samples for 3 min at 1500 × *g* at 4 °C. Remove the supernatant with a 200 μL pipette. Avoid disturbing the beads when discarding supernatant. Wash beads 3× with 1 mL TE buffer (with 1× protease inhibitor cocktail). During each washing, place the samples on a rotating mixer for 10 min at 4 °C. Pellet

beads after the last TE wash, resuspend them at 1:1 volume ratio with TE buffer (with 1× protease inhibitor cocktail) and store at 4 °C (*see* **Note 5**).

13. Pre-clean the samples from **step 10** by adding 50 μL of protein-A/G 50% slurry from the previous step, then rotate samples on a rotating mixer for 2 h at 4 °C.
14. Pellet the beads by centrifuging samples for 3 min at 2500 × *g* at 4 °C.
15. Equally, distribute the supernatant (~300 μL) to a fresh non-stick centrifuge tube. Add fish sperm DNA and BSA to final concentration 0.025 μg/μL and 0.25 μg/μL, respectively. Immunoprecipitate with 2–5 μg of target antibody or normal IgG as a mock sample. Rotate the samples on a rotating mixer overnight at 4 °C.
16. The following morning add 30 μL of protein-A/G Sepharose slurry and continue incubating the samples on a rotating mixer for 2 more hours at 4 °C.
17. Wash the precipitates sequentially with 1 mL each of TSE I, TSEII and LiCl. Wash 2× with TE for 10 min each wash at room temperature on a rotating mixer. Transfer samples to a new microcentrifuge tube after the first TE wash. Pellet the beads by centrifugation for 5 min at 2500 × *g* at 4 °C. Remove any traces of supernatant carefully.
18. Add 120 μL of freshly prepared TE with 1% SDS to each sample and vortex mix. After centrifuging the tubes briefly, incubate the samples at 65 °C for 4 h or overnight to reverse the crosslink process and elute the protein-DNA complex from the sepharose beads.
19. Centrifuge at 2500 × *g* for 3 min and transfer the supernatant to a new microcentrifuge tube.
20. Purify the DNA using Qiagen PCR clean-up kit by adding 600 mL PB buffer to the sample, and following the manufacturer's protocol. We recommend washing the sample twice with the PE buffer.
21. Elute DNA with 40 μL EB buffer from the Qiagen PCR clean-up kit.

3.2 Amplification and Labeling of ChIPed DNA for the Hybridization to Affymetrix Tiling Microarray

This section is the working protocol based on the Affymetrix® chromatin immunoprecipitation assay protocol [14]. Please find the details about the buffers and reagents in the manufacture's protocol.

1. Set up the linear amplification reaction by using ChIPed DNA with target antibody or IgG in a PCR tube as the following reaction. One amplification reaction of a single array. Three reactions are needed for the multi-array set. Each reaction

consists of 10 μL ChIPed DNA, 4 μL 5× Sequenase™ Reaction buffer, 4 μL Primer A (200 μM).

2. Mix, centrifuge tube and incubate at 95 °C for 4 min in a PCR machine. Then after snap cooling samples on ice for 3 min, incubate samples at 10 °C with the PCR machine.
3. Prepare the first round cocktail as the following reaction: 0.1 μL 20 mg/mL BSA, 1 μL 0.1 M DTT, 0.5 μL 25 mM dNTPs, 1 μL Diluted Sequenase™ (1/10 from 13 U/μL stock).
4. Add 2.6 μL above cocktail to each reaction and mix well by pipetting. Run the PCR as the following program: 10 °C 5 min, ramp from 10 °C to 37 °C over 9 min, 37 °C 8 min, 95 °C 4 min.
5. Snap cool samples on ice for 3 min, then incubate samples at 10 °C with the PCR machine.
6. Add 1.0 μL of 1.3 U/μL Sequenase™ to each sample, continue to run PCR machine with the following program: 10 °C 5 min, ramp from 10 °C to 37 °C over 9 min, 37 °C 8 min, 95 °C 4 min.
7. Repeat procedures in **steps 5** and **6** two more times. Then incubate samples at 4 °C.
8. Purify the DNA using Qiagen PCR clean-up kit by adding 120 μL PB buffer to the sample following the manuscript's protocol. Then elute DNA with 40 μL EB buffer from the Qiagen kit.
9. Continue to set up the PCR amplification reaction using the purified DNA above as the following: 20 μL DNA purified for **step 8**, 10 μL 10× PCR buffer, 3.75 μL dNTPs + dUTP, 4 μL 100 μM primer B, 2 μL 5 U/μL Taq Polymerase, 60.25 μL nuclease-free water.
10. Run PCR as the following program: 15 cycles of (95 °C 30 s, 45 °C 30 s, 55 °C 30 s, 72 °C 1 min), then 15 cycles of (95 °C 30 s, 45 °C 30 s, 55 °C 30 s, 72 °C 1 min (for every subsequent cycle add 5 s to this step)), hold at 4 °C.
11. Purify the DNA using Qiagen PCR clean-up kit by mixing 600 μL PB buffer with the sample following the manufacturer's protocol and elute DNA with 40 μL EB buffer from the Qiagen kit.
12. Measure DNA concentration using a NanoDrop or other UV-vis spectrophotometer. Normally, more than 9 μg of amplified DNA is obtained from each reaction.
13. Fragment the samples using the reaction below for a multi-array set: 9.0 μg dscDNA, 4.8 μL 10× APE 1 reaction buffer,

4.0 μL Uracil-DNA Glycosylase (UDG) (2 U/μL), 7.0 μL APE 1 (10 U/μL), up to 32.2 μL nuclease-free water.

14. Mix the sample well and incubate at 37 °C 1 h. Then heat to 93 °C for 2 min and incubate at 4 °C for at least 2 min.
15. Prepare the double-stranded DNA labeling mix as described below: 12 μL 5× TdT Reaction Buffer, 2 μL rTdT 30 U/μL, 1 μL DNA Labeling Reagent 10 mM.
16. Mix 15 μL of the double-stranded DNA labeling mix and 45 μL of the DNA sample from **step 14**. Repeat **step 14**. The labeled DNA is ready for microarray hybridization.

4 Notes

1. Since ChIP is based on immunoprecipitation, the specificity of the antibody used and the degree of enrichment achieved through the affinity precipitation step are crucial for the quality of a ChIP experiment. A very detailed protocol for selecting and validating the antibody for ChIP can be found in the guidelines for ChIP from ENCODE [12] and elsewhere [16].
2. The most commonly used crosslinking reagent for ChIP is formaldehyde (HCHO). It is a powerful and easily reversible cross-linking agent with a short cross-linking spacer arm of 2 Å. Thus it is the most useful reagent to examine proteins directly bound to DNA, such as transcription factors, histones, and their covalent modifications [17]. Fixation conditions need to be optimized for different cell types. Excessive cross-linking results in a substantial loss of material and a higher average of DNA fragments after sonication [17]. The 10 min room temperature fixation in this protocol works well for most cell types in our hands.
3. Sonicate the samples to lyse the pellets and shear the DNA to an average of 500–1000 bp fragments. The time and number of pulses required to shear DNA depend on the cell type, extent of cross-linking, and the instrument used for sonication. For most adherent cells, 10 cycles of sonication using microtip at power 2.5 with 90% duty for 10 s in a Branson Sonifier 250 is sufficient. It is recommended to optimize the sonication conditions by performing a sonication time course and analyzing the products on a 1% agarose gel to identify the optimal conditions that result in an appropriate size range of fragments. Optimal shearing will result in DNA ranging from 0.3 to 2 kb in size with most of the fragments concentrated in the 0.5–1 kb size range.
4. Triton X-100 is a mild nonionic detergent. It binds most water-soluble proteins at only a few high-affinity sites and this binding almost never results in denaturant conformational changes in

the proteins. Thus the higher concentration of Triton X-100 will facilitate the solubilization of proteins without strongly affecting either the antibody binding or protein complex formation occurring during immunoprecipitation.

5. The ready-to-use bead slurry could be stored at 4 °C for up to 1 week.

References

1. Ho JW, Bishop E, Karchenko PV et al (2011) ChIP-chip versus ChIP-seq: lessons for experimental design and data analysis. BMC Genomics 12:134. doi:10.1186/1471-2164-12-134
2. Carroll JS, Meyer CA, Song J et al (2006) Genome-wide analysis of estrogen receptor binding sites. Nat Genet 38(11):1289–1297. doi:10.1038/ng1901
3. Siersbaek R, Nielsen R, John S et al (2011) Extensive chromatin remodelling and establishment of transcription factor 'hotspots' during early adipogenesis. EMBO J 30 (8):1459–1472. doi:10.1038/emboj.2011.65
4. Pellegrini M, Ferrari R (2012) Epigenetic analysis: ChIP-chip and ChIP-seq. Methods Mol Biol 802:377–387. doi:10.1007/978-1-61779-400-1_25
5. Graf S, Nielsen FG, Kurtz S et al (2007) Optimized design and assessment of whole genome tiling arrays. Bioinformatics 23(13):i195–i204. doi:10.1093/bioinformatics/btm200
6. Buck MJ, Lieb JD (2004) ChIP-chip: considerations for the design, analysis, and application of genome-wide chromatin immunoprecipitation experiments. Genomics 83(3):349–360
7. Carey MF, Peterson CL, Smale ST (2009) Chromatin immunoprecipitation (ChIP). Cold Spring Harb Protoc 2009(9):pdb prot5279. doi:10.1101/pdb.prot5279
8. Gobel U, Reimer J, Turck F (2010) Genome-wide mapping of protein-DNA interaction by chromatin immunoprecipitation and DNA microarray hybridization (ChIP-chip). Part B: ChIP-chip data analysis. Methods Mol Biol 631:161–184. doi:10.1007/978-1-60761-646-7_13
9. Reimer JJ, Turck F (2010) Genome-wide mapping of protein-DNA interaction by chromatin immunoprecipitation and DNA microarray hybridization (ChIP-chip). Part A: ChIP-chip molecular methods. Methods Mol Biol 631:139–160. doi:10.1007/978-1-60761-646-7_12
10. Tong Y, Falk J (2009) Genome-wide analysis for protein-DNA interaction: ChIP-chip. Methods Mol Biol 590:235–251. doi:10.1007/978-1-60327-378-7_15
11. Openwetware (2006) ChIP protocol. http://openwetware.org/wiki/ChIP
12. Landt SG, Marinov GK, Kundaje A et al (2012) ChIP-seq guidelines and practices of the ENCODE and modENCODE consortia. Genome Res 22(9):1813–1831. doi:10.1101/gr.136184.111
13. Abcam Beginner's guide to ChIP presented at Abcam. http://docs.abcam.com/pdf/chromatin/A-beginners-guide-to-ChIP.pdf
14. Affymetrix (2005) Affymetrix® chromatin immunoprecipitation assay protocol. http://media.affymetrix.com/support/downloads/manuals/chip_702238_protocol_manual.pdf
15. Yoder SJ (2015) ChIP-on-chip analysis methods for Affymetrix tiling arrays. Methods Mol Biol 1288:473–487. doi:10.1007/978-1-4939-2474-5_27
16. Wardle FC, Tan H (2015) A ChIP on the shoulder? Chromatin immunoprecipitation and validation strategies for ChIP antibodies. F1000Res 4:235. doi:10.12688/f1000research.6719.1
17. Orlando V, Strutt H, Paro R (1997) Analysis of chromatin structure by in vivo formaldehyde cross-linking. Methods 11(2):205–214. doi:10.1006/meth.1996.0407

Chapter 5

Chromatin Immunoprecipitation from Mouse Embryonic Tissue or Adherent Cells in Culture, Followed by Next-Generation Sequencing

Mário A.F. Soares and Diogo S. Castro

Abstract

Chromatin immunoprecipitation (ChIP) is considered the method of choice for characterizing interactions between a protein of interest and specific genomic regions. It is of paramount importance in gene-regulation studies, as it can be used to map the target regions of sequence-specific transcription factors and cofactors, or histone marks that characterize distinct chromatin states. ChIP can be used directly to probe interactions with candidate regions (ChIP-PCR), or coupled to Next-Generation Sequencing (ChIP-seq) to generate genome-wide information. This chapter describes a protocol for performing ChIP and ChIP-seq of transcription factors, starting either from mouse embryonic tissue or adherent cells in culture.

Key words ChIP, ChIP-seq, Chromatin, Mouse embryo, Neural stem/progenitor cells

1 Introduction

Chromatin immunoprecipitation (ChIP) is the obvious technique to study the interaction between a protein of interest and specific genomic regions within its native cellular environment. This technique is most frequently used to map the binding sites of transcription factors or the location of modified histone proteins, but can be applied essentially to any other protein that comes in close vicinity to DNA. In a ChIP assay proteins are crosslinked to DNA using a fixation reagent applied directly to living cells, and chromatin is extracted and sheared into small fragments. The protein of interest (together with crosslinked DNA) is immunoprecipitated with a specific antibody, crosslinks are reversed and the DNA purified. The end result of a ChIP assay is a pool of DNA fragments that is highly enriched for the genomic regions bound by the protein of interest. This can be used for assessing the interactions with candidate sites by quantifying the relative abundance of various genomic regions (ChIP-PCR). Alternatively, precipitated DNA can be

Neus Visa and Antonio Jordán-Pla (eds.), *Chromatin Immunoprecipitation: Methods and Protocols*,
Methods in Molecular Biology, vol. 1689, DOI 10.1007/978-1-4939-7380-4_5, © Springer Science+Business Media LLC 2018

subjected to Next-Generation Sequencing (ChIP-seq), giving rise to genome-wide occupancy profiles [1, 2].

In this chapter, we describe a ChIP protocol we have used extensively for mapping the binding sites of sequence-specific transcription factors expressed in neural progenitor cells [3–6]. It includes the procedures used for crosslinking and extracting chromatin from mouse embryonic brain tissue (in particular dorsal or ventral telencephalon), or from adherent cultures of neural stem/progenitor cells [7]. This same protocol should be applicable to other embryonic tissues and/or developmental stages, or to most cell types that grow in adherent conditions, provided some steps are optimized (discussed in Subheading 4).

Two types of fixation (single or double) have been mostly described in ChIP protocols. Single fixation with formaldehyde (FA) crosslinks methylene bridges, linking proteins to DNA in a reversible manner. In this protocol, we use a double fixation method initially described by Tian and colleagues [8] which in addition to FA requires a second fixation reagentDouble fixation) (disuccinimidyl glutarate or DSG). This results in crosslinking between proteins over long distances (7.7 Å as opposed to 2 Å with FA) via NHS esters. Although double fixation is particularly suited to study proteins that are indirectly recruited to DNA via protein–protein interactions (e.g., transcriptional cofactors), we recommend using this type of fixation also when performing location analysis of sequence-specific transcription factors.

As opposed to some other protocols available, we perform chromatin extraction without a nuclear isolation step. This is of particular relevance when dealing with minute amounts of embryonic tissue, so that pooling of material from various rounds of dissections is not required. Shearing of chromatin is achieved using a Bioruptor, a water-bath sonicator specially conceived for chromatin work.

At the end of the ChIP protocol, immunoprecipitated material may be used for testing the enrichment of candidate genomic regions by ChIP-qPCR, or for de novo identification of target regions on a genome-wide scale by ChIP-seq. In the second case, qPCR on a small fraction of the immunoprecipitated material must also be performed, to confirm ChIP efficiency before more costly procedures associated with NGS are undertaken. To prepare a ChIP-seq sample, we recommend pooling the product from multiple ChIP reactions, as opposed to scaling-up the ChIP reaction.

2 Materials

Use only molecular biology grade reagents. Solutions should be prepared with autoclaved Milli-Q water (or nuclease-free water when indicated).

2.1 General Materials and Equipment

1. Adherent cultures of neural stem/progenitor cells or mouse embryos at E12.5 stage of development.
2. Cell scrapers.
3. Clear, nuclease-free, and non-sticky Eppendorf tubes (1.5 mL).
4. Conical tubes (15 mL).
5. Tabletop centrifuges (for conical and Eppendorf tubes, with temperature control).
6. Tube rotator (for Eppendorf tubes, both in cold room and at RT).
7. Magnetic rack (suitable for Eppendorf tubes).
8. Heat block (for Eppendorf tubes).
9. Micropipettes with filter tips.
10. Agarose gel electrophoresis apparatus.
11. Portable container for liquid N_2.
12. Water-bath sonicator with cooling unit (Bioruptor or similar).
13. Absorbance spectrophotometer (Nanodrop or similar).
14. Real-time PCR machine and suitable multi-well plates.

2.2 Crosslinking and Chromatin Extraction

1. Phosphate buffered saline (PBS).
2. DSG, 5 mg or 50 mg vials (Sigma).
3. Dimethyl sulfoxide (DMSO).
4. Formaldehyde, 37% (FA).
5. Glycine, 2.5 M stock solution.
6. Protease inhibitor cocktail tablets.
7. SDS lysis buffer: 50 mM Tris–HCl pH 8.0, 10 mM EDTA, 1% SDS (add protease inhibitors before use).
8. Sodium Chloride (NaCl), 5 M stock solution.
9. Proteinase K, 10 mg/mL stock solution.
10. Phenol/chloroform, previously equilibrated with TE (pH 8.0).
11. Glycogen, 20 μg/μL.
12. 3 M Sodium Acetate (CH_3COONa) pH 5.2.
13. Ethanol >99.5%.
14. Agarose for analytical gel electrophoresis.
15. Isopropanol.
16. Chloroform:Isoamyl alcohol (24:1).
17. Nuclease-free water.

2.3 Chromatin Immunoprecipitation

1. Antibodies ChIP-grade.
2. Protein G-coated magnetic beads.
3. 10× IP buffer: 0.2 M HEPES pH 8.0, 20 mM EDTA, 2 M NaCl.
4. IP buffer: dilute from 10× stock, completing with 0.1% Sodium Deoxycholate (Na-DOC), 1% Triton X-100, 1 mg/mL BSA and protease inhibitors.
5. LiCl Washing buffer: 50 mM HEPES pH 7.6, 1 mM EDTA, 1% NP-40, 0.7% Na-DOC, 0.5 M LiCl.
6. NaCl Washing buffer: 50 mM HEPES pH 7.6, 1 mM EDTA, 1% NP-40, 0.7% Na-DOC, 1 M NaCl.
7. TE buffer: 10 mM Tris–HCl pH 8.0, 1 mM EDTA.
8. Elution buffer: 50 mM Tris–HCl pH 8.0, 10 mM EDTA, 1% SDS.

3 Methods

3.1 Chromatin Extraction from Embryonic Mouse Brain

This section describes the extraction of chromatin from either dorsal or ventral telencephalon of E12.5 mouse embryos, assuming the reader is familiar with the dissection procedure. Protocol can be applied to other embryonic brain regions (and stages of development), provided cross-linking and sonication conditions are adjusted.

1. Dissect out telencephalons in ice-cold PBS, maintaining the dissected pieces on ice. Separate dorsal and ventral regions and collect the pool of each tissue type in 1 mL PBS in an Eppendorf tube. Protocol is designed for starting material from 18 embryos.
2. Prepare the DSG fixative by adding 30.6 μL DMSO to a 5 mg vial before use (*see* **Note 1**).
3. Add 4 μL DSG fixative to the tissue in PBS (2 mM final concentration), and incubate for 20 min at RT while rotating (*see* **Note 2**).
4. Spin down for 1 min at 850 RCFs (approximately 3000 RPM in a tabletop centrifuge). Wash twice by loosening the tissue in 1 mL of PBS containing Protease Inhibitors (PIs).
5. Keep tissue in 1 mL PBS containing PIs and add 27 μL FA (1% final concentration). Rotate for 10 min at RT.
6. Quench the fixation reaction by adding 50 μL 2.5 M Glycine (125 mM final concentration) for 5 min at RT while rotating.
7. Spin down for 1 min at 850 RCFs at 4 °C. Aspirate supernatant carefully and wash 3 times with 1 mL PBS containing PIs.

From this point on, use nuclease-free, non-sticky Eppendorf tubes.

8. Lyse the tissue in SDS Lysis buffer containing PIs (dorsal tissue in 600 μL, ventral in 900 μL). Pipette up and down with a 200 μL tip to roughly dissociate the tissue. Rotate for 20 min at 4 °C.
9. While waiting, pre-cool sonicator to 4 °C. Set the machine to 200 W (high) and 30 s ON/30 s OFF for 42 min (*see* **Note 3**).
10. Aliquot the chromatin lysate in 300 μL aliquots in Eppendorf tubes and proceed with sonication. Use empty Eppendorf tubes to fill in any empty space in the water-bath, to keep conditions constant across experiments (*see* **Note 4**).
11. Pool the content of different tubes and spin down at maximum speed for 10 min at 4 °C (*see* **Note 5**).
12. Save supernatant and measure the A260 to determine concentration (expect 0.6–1.0 μg/μL). Aliquot conveniently and snap freeze in liquid N_2. Leave aside a 10 μL aliquot to assess sonication efficiency (*see* Subheading 3.3).

3.2 Chromatin Extraction from Adherent Cultures of Neural Stem/Progenitor Cells

This section describes the extraction of chromatin from adherent cultures of neural stem/progenitor cells. The principles followed are similar to the ones described for chromatin preparation from embryonic mouse brain tissue, and it should be applicable to other cell types grown in adherent conditions.

1. Remove medium and wash cells twice with PBS, leaving 17.5 mL in each T-150 flask after last wash. We recommend starting with 4× T-150 confluent flasks (each flask containing approximately 32 million cells).
2. Prepare the fixative by adding 306 μL DMSO to a vial of 50 mg of DSG before use (*see* **Note 1**).
3. Add 70 μL of DSG solution to the 17.5 mL PBS contained in each flask (2 mM final concentration) and incubate at bench only with occasional agitation for 45 min at RT (*see* **Note 6**).
4. Pour off PBS/DSG solution and wash carefully twice with PBS. Leave 20 mL PBS after last wash.
5. Add 540 μL FA (1% final concentration) and incubate for 10 min at RT.
6. Quench reaction by adding 1 mL of Glycine (125 mM final concentration) and incubate for 5 min at RT.
7. Wash cells twice with PBS, leaving 7.5 mL PBS containing protease inhibitors in each flask.
8. Use cell-scraper to detach cells, and pool content of every two flasks into one 15 mL conical tube. Spin-down cells at 400

RCFs (approximately 1300 RPM in a tabletop centrifuge) for 7 min at 4 °C.

9. Carefully remove supernatant and lyse cell pellet in SDS Lysis buffer (use about 300 μL lysis buffer per T-150 flask of confluent cells and incubate for 20 min at 4 °C. From this point on, use nuclease-free, non-sticky Eppendorf tubes.
10. Sonicate lysate in 300 μL aliquots as described in previous section, applying 15 sonication cycles instead of 42 (*see* **Notes 3** and **4**).
11. Pool the content of different tubes and spin down at maximum speed for 10 min at 4 °C (*see* **Note 5**).
12. Save supernatant and measure the A260 to determine concentration (expect 0.6–1.0 μg/μL). Aliquot conveniently and snap freeze in liquid N_2. Leave aside a 10 μL aliquot to assess sonication efficiency (*see* Subheading 3.3).

3.3 Assess Efficiency of Sonication

To assess the efficiency of sonication after chromatin extraction, crosslinks in a small aliquot of chromatin extract are reversed and the DNA is purified and run on a high percentage agarose gel.

1. To reverse FA crosslinks of the 10 μL aliquot of chromatin extract, add 200 μL of elution buffer, 5 μL NaCl and 2 μL proteinase K. Incubate in a heat-block for 2 h at 42 °C, and subsequently overnight at 65 °C.
2. Remove tubes from heat-block and let them cool-down to RT. Extract once with an equal volume of phenol/chloroform, vortex vigorously and centrifuge at maximum speed for 10 min at RT.
3. Remove the aqueous phase to a new Eppendorf tube. Add 1 μL glycogen, 20 μL Sodium Acetate and vortex. Add 440 μL of ice-cold ethanol, vortex vigorously and incubate at −20 °C for at least 30 min.
4. Centrifuge at maximum speed for 20 min at 4 °C, and dissolve pellet in nuclease-free water in its original volume.
5. Measure the A260 to determine DNA concentration and run 1 μg and 2 μg in separate lanes of a 2% analytical agarose gel (*see* **Note 7**).

3.4 Chromatin Immunoprecipitation Assay

This section of the protocol describes the chromatin immunoprecipitation assay using chromatin extracted from either tissue or cultured cells. Section ends with quantification of genomic regions by quantitative PCR (ChIP-PCR), while the following section describes how to proceed to prepare a ChIP-seq sample.

1. Equilibrate the total amount of beads to be used by washing beads 4 × 1 mL of IP buffer. Resuspend beads in IP buffer in its original volume (*see* **Note 8**).
2. If necessary, start by diluting the chromatin stock in lysis buffer, so that the amount of chromatin used per ChIP reaction is contained in 100 μL (*see* **Note 9**). Set aside a small aliquot corresponding to 5% of input chromatin of a ChIP reaction (for final PCR).
3. Dilute the chromatin with twice the volume of IP buffer, and pre-clear it with 50 μL of beads, rotating for >1.5 h (up to 4 h) at 4 °C.
4. Set an Eppendorf tube for each IP reaction, adding 700 μL of IP buffer and the antibody (*see* **Note 10**). Prepare a negative control ChIP reaction tube (e.g., Mock) (*see* **Note 11**).
5. Capture beads with magnetic rack and pipette 300 μL of pre-cleared chromatin to each ChIP reaction. Let tubes rotate overnight at 4 °C.
6. Add 50 μL of beads and rotate for 2 h at 4 °C.
7. Capture beads with magnetic rack, and wash them sequentially 3× LiCl washing buffer, 2× NaCl washing buffer and 1× TE buffer (1 mL each wash). Rotate tubes for 4 min at RT during each wash (*see* **Note 12**).
8. Resuspend beads in 500 μL of Elution buffer and top up the 5% input tube with Elution buffer up to 500 μL (all tubes are treated equally from this point). Incubate for 12 min at 65 °C (vortex briefly before and after incubation).
9. Capture beads and transfer each eluted fraction to a new tube (*see* **Note 13**). Discard beads.
10. Add 5 μL of Proteinase K and 11 μL of NaCl. Vortex and incubate for 2 h at 42 °C.
11. Transfer tubes to 65 °C and incubate overnight to reverse FA crosslinks (*see* **Note 14**).
12. Remove tubes from heat-block and let them cool-down to RT. Extract once with 500 μL phenol/chloroform and once with 500 μL chloroform:isoamyl alcohol, retaining the top aqueous phase (in each extraction step vortex vigorously before centrifuging at maximum speed for 10 min at RT to separate phases).
13. Add 2 μL glycogen, 50 μL of sodium acetate and vortex.
14. Add 0.9 mL of isopropanol and vortex vigorously for 30 s, placing immediately each tube on ice. Transfer all tubes to −20 °C for at least 20 min.
15. Centrifuge for 20 min at full speed at 4 °C.

16. Discard supernatant, and wash once carefully with at least 200 μL of 70% Ethanol (*see* **Note 15**).
17. Air-dry pellet at RT or in heat-block at 42 °C until it becomes transparent and dissolve it in 120 μL nuclease-free water (*see* **Note 16**).
18. Proceed with real-time PCR, using 5 μL of ChIPed DNA per reaction (*see* **Note 17**).

3.5 Preparation of ChIP-seq Samples

Perform ChIP reactions as previously described, and pool precipitated material by dissolving various pellets in a small volume (typically 40 μL) of water. The number of ChIP reactions required may vary greatly depending on antibodies used. Use only 1 μL glycogen per ChIP reaction to decrease its concentration in final sample. Dilute a small aliquot of this into a larger volume of water to test the enrichment of positive control regions by PCR. Typically we quantify two positive and two negative regions for quality control purposes. If antibody was previously tested by ChIP-qPCR, we do not usually prepare a negative control ChIP reaction when preparing a ChIP-seq sample. Final concentration of sample must be determined by fluorescence (e.g. Qubit), before sending for sequencing.

4 Notes

1. Dissolve DSG in DMSO to a 0.5 M concentration short before use (do not save any leftover).
2. When starting from embryonic tissue from another region or developmental stage, crosslinking conditions compatible with efficient sonication may need to be optimized (start by changing incubation times with fixation reagents).
3. Settings provided were optimized to the indicated starting material. New settings should be determined empirically (typically varying the number of cycles) when starting with tissue from another embryonic region or cell type. Carefully follow the Bioruptor manufacturer's protocol so as not to heat-up the device, which may require pausing for a short period of time after a certain number of cycles.
4. SDS in lysis buffer may precipitate if kept on ice for a long period of time. Make sure all SDS is in solution when in water-bath, as otherwise sonication will not be efficient.
5. After sonication, chromatin preparation should appear clear and only a minor pellet may be found after centrifugation.
6. A white precipitate will always form when adding DSG to the PBS. Add it very slowly, while gently mixing, to avoid major

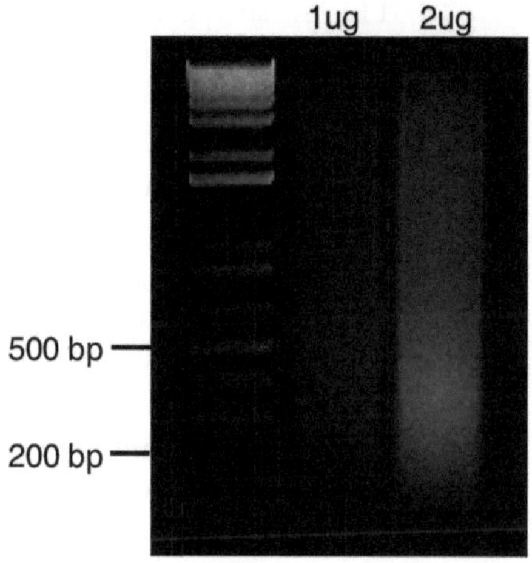

Fig. 1 Chromatin extracted from adherent cultures of neural stem/progenitor cells and run on a 2% agarose gel upon reversal of crosslinks. Most fragments are within 200–500 bp

precipitation. Begin by tilting the flask so that the precipitate will not fall on cells, and agitate gently to dissolve it.

7. Run two different amounts of chromatin in separate lanes, to make sure at least one lane is not overloaded. Most DNA fragments should be within the 200 bp and 500 bp size window (Fig. 1).
8. We strongly recommend using magnetic beads. Protein G beads are a preferred choice for human and mouse samples since this type of protein binds with more affinity, and to a wider range of IgG subtypes when compared to Protein A. Nevertheless, the type of coated-beads must be chosen according to the antibody. Use 50 μL beads per sample. Do not forget to include required amount of beads for pre-clearing chromatin.
9. For ChIPing transcription factors, use at least 30 μg of chromatin per ChIP-reaction. To obtain reproducible results, it is very important to keep the concentration of SDS constant at 0.1% in the ChIP reaction by diluting 1:10 the chromatin stock.
10. Not all antibodies are suitable for ChIP. In some cases, suppliers may refer to an antibody as being ChIP-grade. Because it is difficult to anticipate if a particular antibody will work, it is very important to design as many positive control regions as possible (*see* **Note 17**), in order to interpret any negative results. The optimal amount of antibody to use may be determined empirically by testing various quantities against a constant amount of beads (0.2–2 μg).

11. An isotype control IgG may be used as a negative control. Instead, we routinely use a mock ChIP reaction where no antibody is used.
12. The number and sequence of washes may be optimized for each antibody.
13. Debris from degradation of beads may appear after incubation at 65 °C. These will not be a problem and may be removed with a quick spin, and will be eventually lost during phenol/chloroform extraction.
14. Remove tubes as early as possible after overnight incubation, as longer incubation times are not desirable (4 h are sufficient to reverse FA crosslinks).
15. Wash carefully as pellets will eventually become loose and tend to float (do not spin down again as this may break them). Instead, remove as much volume as possible with pipette, leaving a small drop.
16. Surveil tubes if drying pellets at 42 °C and dissolve them while they are still wet. The volume suggested is a conservative

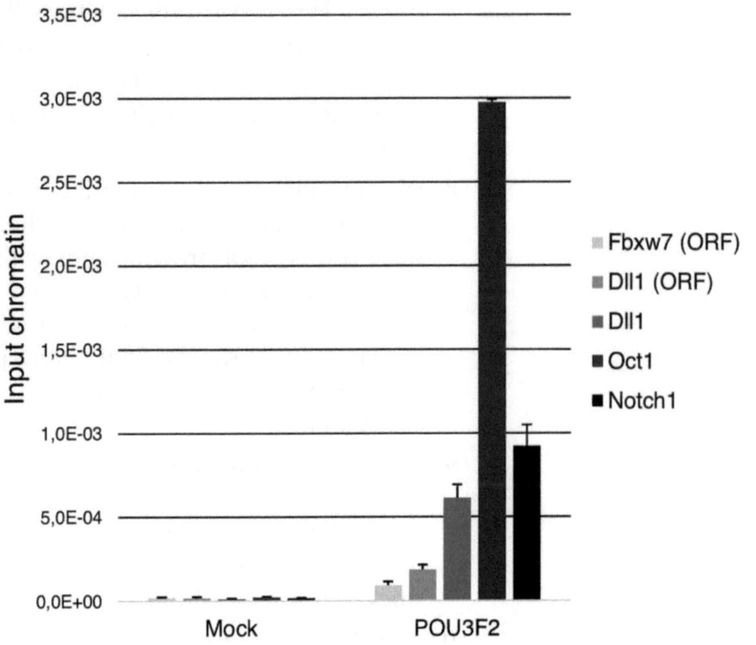

Fig. 2 Real-time PCR on immunoprecipitated material obtained from a ChIP reaction for Pou3f2 and chromatin extracted from cultured neural stem/progenitor cells. Various degrees of enrichment are observed in regulatory regions associated with Dll1, Oct1, and Notch1, when compared to Dll1 ORF and Fbxw7 ORF (negative control regions). No enrichment between regions is found in the absence of antibody (mock). Values are the average of triplicate samples and *error bars* represent the standard deviation

estimate and larger volumes may be used depending on the amount of material precipitated by the antibody.

17. We perform real-time PCR reactions in 20 μL total volume, using SYBR-Green. The result of the ChIP assay is determined by the relative enrichment of bound regions over negative control regions, in an antibody-dependent manner (Fig. 2). Quantifications are done across a standard curve established from serial dilutions of the 5% input chromatin sample. For transcription factors, a negative control region can be located for example within an open reading frame of the putative target gene. For simplification, we use a constant set of negative control regions that are not necessarily located within the loci of candidate target regions being tested, as we find that values associated with different negative control regions do not vary significantly.

References

1. Park PJ (2009) ChIP-seq: advantages and challenges of a maturing technology. Nat Rev Genet 10:669–680. doi:10.1038/nrg2641
2. Furey TS (2012) ChIP-seq and beyond: new and improved methodologies to detect and characterize protein-DNA interactions. Nat Rev Genet 13(12):840–852. doi:10.1038/nrg3306
3. Monteiro CB, Costa MF, Reguenga C et al (2014) Paired related homeobox protein-like 1 (Prrxl1) controls its own expression by a transcriptional autorepression mechanism. FEBS Lett 588:3475–3482. doi:10.1016/j.febslet.2014.08.006
4. Mateo JL, Van den Berg DLC, Haeussler M et al (2015) Characterization of the neural stem cell gene regulatory network identifies OLIG2 as a multifunctional regulator of self-renewal. Genome Res 25:41–56. doi:10.1101/gr.173435.114
5. Raposo AA, Vasconcelos FF, Drechsel D et al (2015) Ascl1 coordinately regulates gene expression and the chromatin landscape during neurogenesis. Cell Rep 10:1544–1556. doi:10.1016/j.celrep.2015.02.025
6. Vasconcelos FF, Sessa A, Laranjeira C et al (2016) MyT1 counteracts the neural progenitor program to promote vertebrate neurogenesis. Cell Rep 17:469–483. doi:10.1016/j.celrep.2016.09.024
7. Conti L, Pollard SM, Gorba T et al (2005) Niche-independent symmetrical self-renewal of a mammalian tissue stem cell. PLoS Biol 3:1594–1606. doi:10.1371/journal.pbio.0030283
8. Tian B, Yang J, Brasier AR (2012) Two-step crosslinking for analysis of protein-chromatin interactions. Methods Mol Biol 809:105–120. doi:10.1007/978-1-61779-376-9_7

Chapter 6

Chromatin RNA Immunoprecipitation (ChRIP)

Tanmoy Mondal, Santhilal Subhash, and Chandrasekhar Kanduri

Abstract

Researchers have recently had a growing interest in understanding the functional role of long noncoding RNAs (lncRNAs) in chromatin organization. Accumulated evidence suggests lncRNAs could act as interphase molecules between chromatin and chromatin remodelers to define the epigenetic code. However, it is not clear how lncRNAs target chromatin remodelers to specific chromosomal regions in order to establish a functionally distinct epigenetic state of chromatin. We developed and optimized chromatin RNA immunoprecipitation (ChRIP) technology to characterize the lncRNAs associated with active and inactive chromatin compartments. Use of ChRIP to identify chromatin-bound lncRNA will further improve our knowledge regarding the functional role of lncRNAs in establishing epigenetic modifications of chromatin.

Key words Long noncoding RNA, Chromatin, EZH2, H3K27me3, Gene regulation, Epigenetics, Inactive chromatin, Active chromatin

1 Introduction

Chromatin, a complex of RNA, DNA, and protein, acts as an epigenetically regulated framework that fine-tunes gene expression in a spatiotemporal fashion. The most important constituents of chromatin, DNA, and the histone octamer, are subjected to epigenetic modifications by a complex interplay of DNA and chromatin remodelers in response to developmental and environmental cues. Recent investigations have implicated several long noncoding RNAs (lncRNAs) in the targeting of DNA and chromatin remodelers to specific genomic regions [1, 2]. Though the functional role of DNA and chromatin remodelers in defining the epigenetic code of chromatin is well investigated, the mechanism by which they define the epigenetic code in concert with lncRNAs is still not very clear. A recent investigation suggested that some of the repressive chromatin-associated lncRNAs target chromatin remodelers to specific genomic regions via triplex formation with GA-rich sequences [3]. Hence characterization of chromatin-associated lncRNAs in functionally demarcated chromatin compartments

Neus Visa and Antonio Jordán-Pla (eds.), *Chromatin Immunoprecipitation: Methods and Protocols*,
Methods in Molecular Biology, vol. 1689, DOI 10.1007/978-1-4939-7380-4_6, © Springer Science+Business Media LLC 2018

may help us better understand the functional nexus between lncRNAs and epigenetic states of the chromatin.

Previously, sucrose-fractionated chromatin was used to purify RNA enriched in the chromatin fraction [4]. However, the use of this technique made it difficult to purify RNA from different chromatin compartments such as repressive and active chromatin fractions. Thus a suitable procedure is required to isolate RNA from different chromatin compartments. To this end, we developed a variant of the Chromatin immunoprecipitation (ChIP) protocol, Chromatin RNA immunoprecipitation (ChRIP), to profile chromatin-associated RNA transcripts from different chromatin compartments across the genome. Chromatin immunoprecipitation (ChIP) is a versatile technique that has been used to profile histone modifications and transcription factor occupancy at any given genomic region. Following ChIP, we extract DNA and thus it is possible to identify specific genomic regions enriched in a given ChIP experiment by PCR and sequencing. However, in the case of ChRIP, we extract RNA from the immunopurified chromatin and hence we can only profile RNA transcripts associated with total chromatin and chromatin sub-compartments. Previously, using ChRIP assay, we demonstrated that mouse and human Kcnq1ot1/KCNQ1OT1 lncRNAs are enriched in the repressive chromatin fraction [2]. We later modified this procedure by combining photoactivatable ribonucleoside-enhanced crosslinking followed by high-throughput RNA-sequencing to characterize lncRNAs that are associated with different chromatin fractions. Using this modified ChRIP assay, we purified repressive chromatin using antibodies against H3K27me3 and EZH2 and characterized the associated lncRNAs. EZH2, a component of the PRC2 complex, catalyzes trimethylation of lysine 27 of histone H3 (H3K27me3), a repressive histone associated with gene silencing. Hence, we considered lncRNAs as repressive chromatin associated only when they were enriched in both the EZH2 and H3K27me3 purified chromatin fractions. Photoactivatable ribonucleoside-enhanced crosslinking enabled us to identify potential RNA binding protein (RNP) contact points in the repressive chromatin-enriched lncRNAs. These RNP contact points may serve as functional RNA sequences and these functional sequences may help in uncovering the mode of action of repressive chromatin-associated lncRNAs [3]. Thus the ChRIP protocol not only characterizes chromatin-associated RNAs on a global scale in functionally demarcated chromatin compartments but also provides further insight into mechanisms by which chromatin-associated RNA may regulate gene expression.

2 Materials

2.1 Thiouridine and Actinomycin D Treatment

1. RPMI media supplemented with 10% fetal bovine serum (FBS).
2. T175 flask tissue culture flasks.
3. 4-thiouridine (Sigma Aldrich).
4. Actinomycin D (ActD).
5. 1× PBS: 137 mM NaCl, 10 mM phosphate, 2.7 mM Potassium Chloride pH 7.4.
6. 37% Formaldehyde.
7. 1.25 M Glycine.
8. CL1000 UV crosslinker (UVP) or similar.
9. Cell scraper.

2.2 Nuclei Isolation and Chromatin Preparation

1. 5× Nuclei isolation Buffer: 1.28 M sucrose, 40 mM Tris–HCl, pH 7.5, 20 mM $MgCl_2$, 4% TritonX-100. Supplemented with protease inhibitor cocktail (Roche Life Science) and RNasin Ribonuclease inhibitor 50 U/mL (Promega).
2. Nuclear lysis buffer: 0.1% SDS, 0.5% Triton X-100, 20 mM Tris–HCl pH 7.5, 150 mM NaCl. Supplemented with protease inhibitor cocktail and Ribonuclease inhibitor 50 U/mL.
3. DEPC-treated Milli-Q water.
4. Bioruptor sonicator (Diagenode) or similar.

2.3 Preparation of Antibody-Bound Beads for the Immunoprecipitation

1. Protein A magnetic beads.
2. Bead blocking buffer: Nuclear lysis buffer supplemented with 10 mg/mL BSA.
3. Antibodies specific to chromatin marks and chromatin proteins and control Rabbit IgG.
4. Tube rotator.

2.4 Chromatin Immunoprecipitation

1. Tube rotator.
2. Magnetic Rack.
3. Low salt buffer: 0.1% SDS, 1% TritonX-100, 2 mM EDTA, 20 mM Tris–HCl, pH 7.5, 150 mM NaCl.
4. High Salt Buffer: 0.1% SDS, 1% TritonX-100, 2 mM EDTA, 20 mM Tris–HCl, pH 7.5, 500 mM NaCl.
5. LiCl Buffer: 250 mM LiCl, 1% NP40, 1% deoxycholate, 1 mM EDTA, 10 mM Tris–HCl, pH 7.5.
6. TE buffer: 10 mM Tris–HCl, pH 7.5, 0.1 mM EDTA.
   All the buffers in this section are supplemented with protease inhibitor cocktail.

2.5 Purification of Chromatin-Bound RNA and DNA

1. Elution buffer: 100 mM NaCl, 10 mM Tris–HCl, pH 7, 1 mM EDTA, 0.5% SDS.
2. Proteinase K Solution (20 mg/mL).
3. Heat Block.
4. TRIzol (Thermo Fisher Scientific).
5. DNA Phenol Equilibrated with 10 mM Tris–HCl pH 8.
6. Chloroform.
7. RQ1 RNase-Free DNase I and 10× DNase buffer (Promega).

2.6 RNA Sequencing Analysis Tools

1. Sequence alignment: LifeScope for SOLiD platform, Tophat or HISAT for Illumina.
2. Transcriptome reconstruction: Cufflinks.
3. Read quantification: HTSeq or Subread's Feature Counts.
4. Differential enrichment: EdgeR or DESeq.
5. Finding T to C conversion: Freebayes and TransExtract betaV1.2 or PARAlyzer.

3 Methods

Perform all steps in RNase-free condition to minimize the RNA degradation. Key steps of the ChRIP methods are indicated in Fig. 1a.

3.1 4-Thiouridine (4sU) and Actinomycin D (ActD) Treatment

1. We used the ChRIP-seq protocol extensively on BT-549 human breast cancer cells. Here, we are providing various treatment conditions for BT-549 cells and the conditions can be adapted to any adherent cell cultures. Grow BT-549 in RPMI media as a subconfluent culture in T175 flask with average splitting once a week.
2. For 4sU treatment, seed seven to eight million cells in 150 cm culture dish in the morning and allow the cells to settle down for 6–8 h (*see* **Note 1**). In the evening of the same day, add 4sU at a final concentration of 100 μM in cell culture media and incubate overnight (14–16 h).
3. In the morning of the next day, remove the 4sU containing media, and add ActD containing media at a concentration 5 μg/mL. Treat the cells with ActD for 40–45 min to block cellular transcription and to also get rid of nascent transcripts bound to the chromatin.
4. The efficacy of the ActD treatment can be checked by assessing the expression of mRNAs with a short half-life like cFos and cMyc by RT-qPCR in the ActD-treated and -untreated cells. Normally, we observed 50% downregulation in the expression of *c-Myc* and *c-Fos* in the ActD-treated cells compared) to the untreated cells.

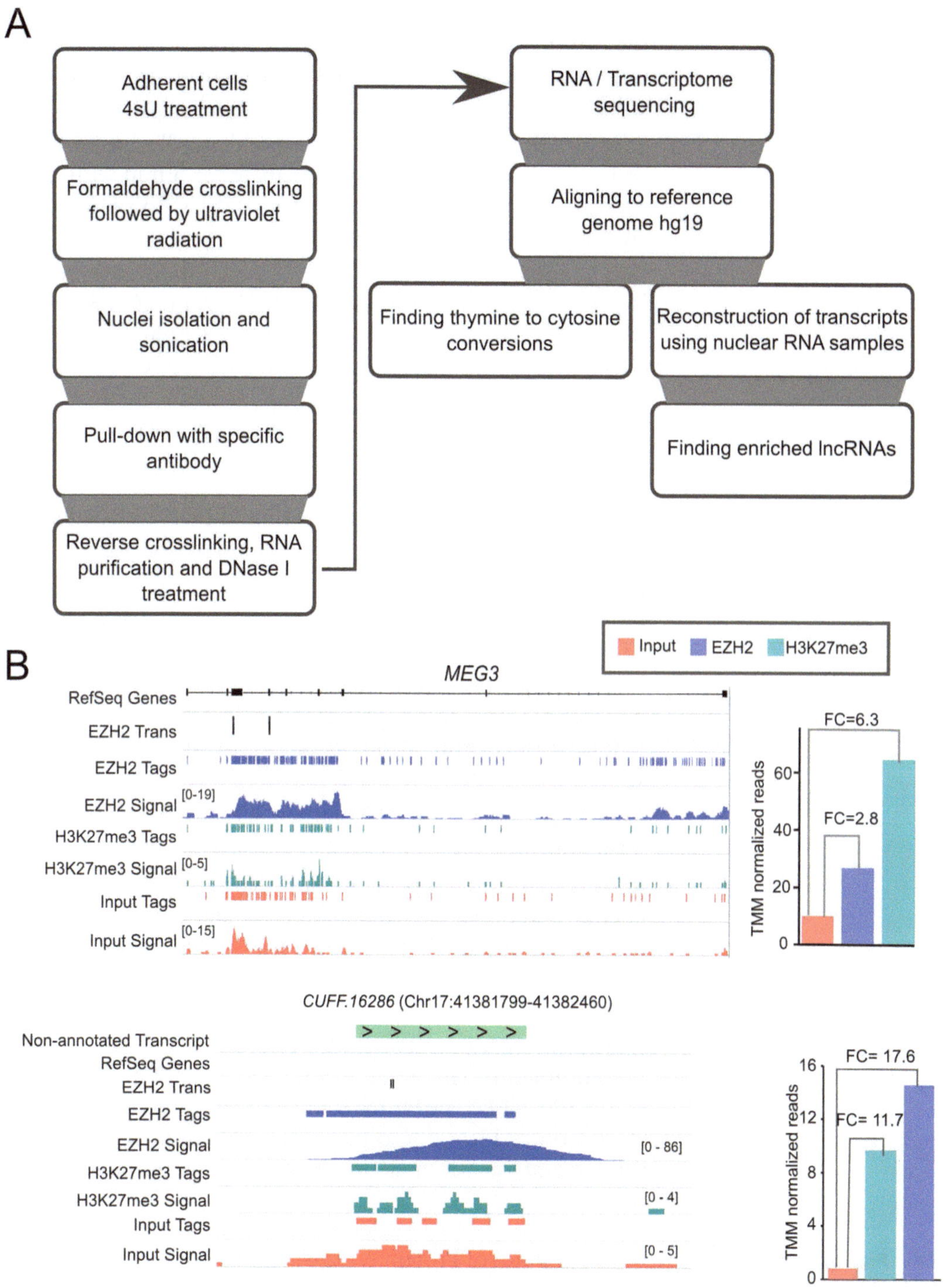

Fig. 1 (**a**) Flow chart depicting the key steps of ChRIP method. (**b**) ChRIP-seq data showing the enrichment of the *MEG3* and a novel lncRNA *CUFF.16286* in EZH2 and H3K27me3 chromatin RNA pull-downs compare to the nuclear input. T to C transitions indicated with *vertical black bars*. Previously published in ref. 3

3.2 Formaldehyde and UV Crosslinking

1. Wash the cells twice with 1× PBS after ActD treatment to remove the residual ActD. Prepare 1% formaldehyde solution in 1× PBS and add 10 mL of 1% formaldehyde per 150 cm dish and keep on rocker for 10 min at 40 rpm.
2. Stop the formaldehyde fixation by adding Glycine to a final concentration of 125 mM and incubate for 5 min with gentle rocking. Wash the cells twice with ice-cold 1× PBS and the cells are ready for UV crosslinking.
3. Add 10 mL of ice-cold 1× PBS to each plate and keep it on ice. Expose the plate to UV irradiation (0.15 J/cm^2 at 365 nm) in a CL1000 cross-linker, or similar. Using a cell scraper, collect the cells into 50 mL falcon tube containing ice-cold 1× PBS. Approximately 10 million cells can be collected from each 150 cm plate and about 20 million cells are needed for each experiment.

3.3 Nuclei Isolation and Chromatin Preparation

For the following steps, the solutions should be pre-chilled on ice prior to their use (*see* **Note 2**).

1. Pellet 20 million UV and formaldehyde crosslinked cells in a 50 mL falcon tube by centrifugation at 0.2 × *g* for 5 min at 4 °C. Resuspend the pellet in 4 mL of ice-cold 1× PBS, 4 mL of 5× nuclei isolation buffer, and 12 mL of DEPC water. Mix the resuspended cells thoroughly by inverting the tube for several times and let the tube stand on ice for 15–20 min (mix the cells at each 5 min interval).
2. Centrifuge at 0.6 × *g* for 10 min at 4 °C to pellet the nuclei and wash once with ice-cold 1× PBS. Discard the supernatant.
3. Resuspend the nuclei pellet in 2 mL Nuclear lysis buffer (1 mL nuclear lysis buffer per 10 million cells) and mix thoroughly to remove any clumps. Transfer the resuspended nuclear pellet to a 1.5 mL tube.
4. Shear the nuclei in nuclear lysis buffer for 20–30 cycles using Bioruptor sonicator (Diagenode) or equivalent to solubilize the chromatin. The average fragment size of the sonicated chromatin is 1 kb (*see* **Notes 3** and **4**). Centrifuge the sonicated chromatin at 11.6 × *g* to pellet the cell debris. The collected supernatant can be used in chromatin immunoprecipitation reactions.

3.4 Preparation of Antibody-Bound Beads for the Immunoprecipitation

1. Transfer 40 μL of Protein A magnetic beads for pull-down with each antibody to 1.5 mL tubes and wash the beads with 500 μL of Nuclear lysis buffer four times. Magnetic beads can be precipitated with magnetic racks by keeping tubes 1 min in close proximity to the magnet until the solution becomes clear and discard the supernatant with 1 mL pipette.

2. Resuspend the beads in 200 μL Bead blocking buffer and rock at 4 °C for 20 min.
3. After blocking, wash the beads twice with Nuclear lysis buffer and resuspend the beads in 300 μL Nuclear lysis buffer.
4. 6 μg of specific antibody against a histone modification of interest or a chromatin modifier can be used in the immunoprecipitation reaction. Previously, we have used antibodies against repressive chromatin-specific histone modification histone H3 lysine 27 trimethylation (H3K27me3) and a chromatin modifier EZH2, which is a part of Polycomb Repressive Complex 2 (PRC2) and catalyzes the H2K27me3 modification. A control antibody without any binding specificity like rabbit IgG can be used. Add antibodies to each 300 μL washed and pre-blocked beads in Nuclear lysis buffer and rock the tubes for 2 h at 4 °C.
5. After 2 h, precipitate the antibody-bound magnetic beads, remove the Nuclear lysis buffer and beads are ready for use in immunoprecipitations (*see* **Note 5**).

3.5 Chromatin Immunoprecipitation

1. For each chromatin RNA immunoprecipitation reaction, take 50–60 μg of soluble chromatin. If required, dilute the chromatin with Nuclear lysis buffer to increase the final volume up to 500 μL. Add the chromatin suspension to antibody-bound beads and rotate the tubes on a rocker overnight at 4 °C (*see* **Notes 6** and 7).
2. For input, 50 μL diluted soluble chromatin can be set aside in a freezer for DNA and RNA extractions the following day.
3. Next day morning, precipitate the magnetic beads using a magnetic rack and remove the supernatant. Wash the beads once with 700 μL of each low-salt and high-salt buffers, followed by two washes of 700 μL LiCl buffer and a single final wash with TE buffer. All the washes can be carried out at 4 °C for 5 min on a rocker.

3.6 Purification of Chromatin-Bound RNA and DNA

1. Resuspend the washed beads from the previous step in 400 μL Elution buffer supplemented with Proteinase K (5 μL of 20 mg/mL). Divide the 400 μL eluate into two 1.5 mL tubes: 300 μL for extracting chromatin-bound RNA and the remaining 100 μL for DNA extraction. At this step, thaw the input samples, divide into two parts for DNA and RNA (25 μL each) and increase their volume to 400 μL with Elution buffer supplemented with Proteinase K (*see* **Note 8**).
2. Incubate at 55 °C 45 min for isolation of chromatin-bound RNA and Input RNA. Incubate 45 °C 4 h for immunoprecipitated DNA and Input DNA extraction. For input RNA samples required for RNA-sequencing *see* **Note 9**.

3. After Proteinase K treatment, reverse the formaldehyde cross-linking by heating the beads from both RNA and DNA extractions at 95 °C 10 min (*see* **Note 10**).
4. Isolate chromatin-bound RNA using standard TRIzol-based extraction, and DNA isolation using standard DNA phenol-chloroform based extraction method. DNA prepared at this step is ready for further use in quantitative PCR (qPCR) for validation of the known binding sites for the histone marks and chromatin modifiers (*see* **Note 11**).
5. To remove DNA contamination from the TRIzol extracted RNA, resuspend the RNA in 40 μL nuclease-free water and add 5 10× DNase I buffer and 5 μL RNase-free DNase I. Mix and incubate at 37 °C for 30 min. After DNase I treatment extract the RNA again using TRIzol, resuspend in 30 μL nuclease-free water and the chromatin-bound RNA is ready for further use depending on the need of the experiment.

3.7 Processing of Chromatin-Bound RNA

1. Chromatin-bound RNA can be used in the standard RT-qPCR reaction to check the enrichment of the RNA of interest. IgG can be used as negative control to verify specific pull-down by any given antibody. mRNAs with no known chromatin interaction can be used to rule out their nonspecific binding in the chromatin pull-downs (*see* **Note 11**). RNA-sequencing can be used to identify chromatin-bound RNA on a global scale, enriched using specific antibodies. To obtain enough amount of chromatin-bound RNA for sequencing, several immunoprecipitation reactions may need to be pooled (*see* **Notes 12–14**).

3.8 Analysis of the RNA Sequencing Data

1. RNA sequencing reads must be cleaned of adaptor sequences before subjecting to alignmentHuman against hg19 genome using RNA-seq splice-aware aligners. The choice of aligners depends on the sequencing platform, for example, LifeScope for color-space SOLiD platform, TopHat [5] or HISAT [6] for Illumina.
2. Transcripts enriched in the samples can be identified using transcriptome reconstruction or assembly with Cufflinks [7], guided by proper transcript annotations like for example Gencode [8] or Ensembl [9].
3. Depending on whether the sequencing reads are stranded or un-stranded, the quantification of reconstructed transcripts can be performed using HTSeq [10] or Subread [11] accordingly (*see* **Note 15**).
4. To evaluate differential enrichment between input and pull-down samples (Fig. 1b), there are many statistical packages available for RNA-seq differential enrichment/expression analysis such as EdgeR [12] and DESeq [13].

5. T to C conversions on binding regions (Fig. 1b) can be identified using haplotype SNP caller such as Freebayes [14] with PERL tweaks [3], (https://github.com/santhilalsubhash/TransExtract_betaV1.2, for strand-specific sequencing protocol) or other specialized algorithms available for PAR-CLIP like PARAlyzer [15] (*see* **Note 16**).

4 Notes

1. If the purpose of the ChRIP experiment is only to check chromatin enrichment on a global scale or a specific subset of the chromatin-associated RNAs, one can avoid the use of 4sU. Use of 4sU in a ChRIP experiment enables subsequent detection of T to C conversion in RNA-seq data. The T to C conversion data have the potential to give rise to probable RNA–protein contact sites. But further experimental validation is required to confirm RNA–protein contact points obtained from the T to C conversions.
2. Keep all buffers supplemented with RNase inhibitor on ice for 15 min prior to the nuclei preparation and sonication.
3. Before starting the main experiment, a pilot experiment can be done to optimize the sonication cycles required for generating 1 kb chromatin fragments. It is also advisable to perform a pilot experiment to assess the yield of the DNA extracted from the sonicated chromatin. In this way, one can directly proceed to immunoprecipitation step on the same day following the sonication step, and this may help to avoid the freeze–thaw cycles of the sonicated chromatin.
4. RNase-free conditions must be strictly ensured throughout the experiment to minimize the degradation of chromatin-bound RNAs.
5. Precipitate antibody-bound beads when the sonicated chromatin is ready for the immunoprecipitation, this will help prevent drying of beads.
6. Try to start the immunoprecipitation reaction late in the day and incubate the reaction overnight. This will prevent long incubation times with antibody while also reducing potential RNA degradation.
7. The amount of antibody required for the immunoprecipitation may vary depending on the quality of the antibody. Standardization of the chromatin pull-downs before starting the ChRIP experiment will enable the use of appropriate amount of antibody.
8. The main purpose of the present protocol is to isolate the chromatin-bound RNA. However, prior to the RNA isolation,

the quality of the chromatin pull-down can be monitored by assessing the enrichment of DNA fragments with known positive controls for a particular antibody. For example, if we use H3K27me3 and EZH2 antibodies for the chromatin pull-downs, we can test the enrichment of DNA fragments which are known to bind to H3K27me3 and EZH2 using qPCR and if the enrichment is verified with the known positive controls, we can then proceed for RNA purification with the immune-precipitated chromatin.

9. For input nuclear RNA isolation a different method should be followed as formaldehyde fixation makes it difficult to isolate intact RNA. After 4sU and ActD treatments as described in the protocol, only UV crosslinking can be performed. Isolate the nuclei and lyse the nuclei in nuclear lysis buffer. Perform Proteinase K treatment as described in the protocol and isolate the RNA using standard TRIzol extraction protocol. Deplete ribosomal RNA using RiboMinus™ Eukaryote System v2 (Life Technologies) and it is ready for RNA-seq.
10. Longer incubation times at 95 °C than the prescribed 10 min during the reverse cross-linking step may lead to degradation of the RNA.
11. To check the enrichment of a particular RNA in a ChRIP experiment, it is advisable to use multiple primers spanning the entire transcript. Due to RNA fragmentation during sonication step, different regions of the RNA may get differentially enriched during the immunoprecipitation step.
12. Pull down the efficiency of an antibody in immunoprecipitation reaction varies depending on the nature of the antibody and this results in differential amount purified RNA from a single immunoprecipitation reaction. So the number of immunoprecipitation reactions that need to be pooled for RNA-seq can be optimized accordingly.
13. If the chromatin-bound RNA needs to be used for RNA-seq, one may need to pool RNA from 6 to 8 chromatin RNA immunoprecipitation experiments, as the RNA recovery from one reaction is not sufficient for RNA-sequencing library preparation using the current library preparation protocols. Pooling of the samples can be done prior to DNase I treatment of the purified chromatin RNA and it is preferable to perform DNase I treatment in 300 μL reaction volume supplemented with 30 μL 10× buffer and 10 μL DNase I.
14. For RNA sequencing (IP, Input, and IgG) any standard RNA-sequencing platform can be used. However, one has to keep in mind that chromatin-bound RNAs are purified after a sonication step and this can lead to RNA fragmentation. Hence one should use a library preparation protocol that does not require

full-length RNA with an intact polyA tail. Previously, we have used SOLiD (Applied Biosystem) RNA-seq platform by following the standard library preparation protocol using the SOLiD Total RNA-Seq Kit.

15. To avoid quantification of the multi-mapped reads, mapping quality filter can be introduced while performing the read quantification.
16. Since the samples are 4sU treated, one can expect thymine to cytosine conversions near the RNA–protein contact sites. Since the interacting regions may have multiple contact sites, multiple conversion sites in the close proximity can be expected. While choosing the conversion sites for further validation, mapping quality of the reads and read depth at conversion sites can be considered.

Acknowledgments

This work was supported by the grants from the Knut and Alice Wallenberg Foundation (KAW) (Dnr KAW 2014.0057), Swedish Foundation for Strategic Research (RB13-0204), Swedish Cancer Research Foundation (Cancerfonden: Kontrakt no. 150796), the Swedish Research Council (VR-M: K2014-67X-20781-07-4), Barncancerfonden (PR2014/0147), Ingabritt och Arne Lundbergs forskningsstiftelse and LUA/ALF to C.K.

References

1. Mohammad F, Pandey GK, Mondal T et al (2012) Long noncoding RNA-mediated maintenance of DNA methylation and transcriptional gene silencing. Development 139 (15):2792–2803
2. Pandey RR, Mondal T et al (2008) Kcnq1ot1 antisense noncoding RNA mediates lineage-specific transcriptional silencing through chromatin-level regulation. Mol Cell 32 (2):232–246
3. Mondal T, Subhash S, Vaid R et al (2015) MEG3 long noncoding RNA regulates the TGF-beta pathway genes through formation of RNA-DNA triplex structures. Nat Commun 6:7743
4. Mondal T, Rasmussen M, Pandey GK et al (2010) Characterization of the RNA content of chromatin. Genome Res 20(7):899–907
5. Kim D, Pertea G, Trapnell C et al (2013) TopHat2: accurate alignment of transcriptomes in the presence of insertions, deletions and gene fusions. Genome Biol 14 (4):R36
6. Kim D, Langmead B, Salzberg SL (2015) HISAT: a fast spliced aligner with low memory requirements. Nat Methods 12(4):357–360
7. Trapnell C, Williams BA, Pertea G et al (2010) Transcript assembly and quantification by RNA-Seq reveals unannotated transcripts and isoform switching during cell differentiation. Nat Biotechnol 28(5):511–515
8. Harrow J, Frankish A, Gonzalez JM et al (2012) GENCODE: the reference human genome annotation for The ENCODE Project. Genome Res 22(9):1760–1774
9. Cunningham F, Amode MR, Barrell D et al (2015) Ensembl 2015. Nucleic Acids Res 43 (Database issue):D662–D669
10. Anders S, Pyl PT, Huber W (2015) HTSeq–a Python framework to work with high-throughput sequencing data. Bioinformatics 31(2):166–169
11. Liao Y, Smyth GK, Shi W (2014) featureCounts: an efficient general purpose program for assigning sequence reads to genomic features. Bioinformatics 30(7):923–930

12. Robinson MD, McCarthy DJ, Smyth GK (2010) edgeR: a Bioconductor package for differential expression analysis of digital gene expression data. Bioinformatics 26(1):139–140
13. Anders S, McCarthy DJ, Chen Y et al (2013) Count-based differential expression analysis of RNA sequencing data using R and Bioconductor. Nat Protoc 8(9):1765–1786
14. Genomes Project Consortium, Abecasis GR, Auton A et al (2012) An integrated map of genetic variation from 1,092 human genomes. Nature 491(7422):56–65
15. Corcoran DL, Georgiev S, Mukherjee N et al (2011) PARalyzer: definition of RNA binding sites from PAR-CLIP short-read sequence data. Genome Biol 12(8):R79

Chapter 7

DNA Accessibility by MNase Digestions

Ann-Kristin Östlund Farrants

Abstract

Micrococcal nuclease (MNase) digestion of chromatin cuts linker DNA between neighboring nucleosomes and in this way generates mononucleosomes. The protected fragments can then be analyzed by genome-wide sequencing techniques or by quantitative PCR to obtain information about the positions of nucleosomes in the chromatin. Nucleosomes are differentially sensitive to MNase digestion, which means that titrations of MNase should be performed to obtain a comprehensive map of the nucleosome positions of a chromatin region or genome.

Key words MNase digestion, Nucleosome position, Chromatin accessibility

1 Introduction

Micrococcal nuclease (MNase) digestion is often used to assess the accessibility of the DNA in protein–DNA complexes, such as chromatin. The tight interactions between histones and DNA in nucleosomes protect the DNA from digestion by MNase, resulting in cuts in the linker DNA between neighboring nucleosomes [10]. Treatment of chromatin with high amounts of MNase generates mononucleosomes, which has given the nucleosome positions in specific regions [11]. Genome-wide sequencing of the DNA fragments has also been used to determine the global profile of nucleosome positions in eukaryote cells (for instance 5, 8). However, some nucleosomes are more sensitive to MNase digestion than others and limited digestion generates other nucleosome profiles than those generated by high amounts of MNase [2, 16]. The nucleosome profile depends on the degree of accessibility of different chromatin regions, and highly transcribed loci are generally more sensitive to MNase digestions because of their more open chromatin structure. Thus, limited MNase digestions can be also used to assess different chromatin states, and global genome-wide nucleosome positioning studies require titrations of MNase in order to obtain a complete nucleosome profile for a given cell type under certain conditions

Neus Visa and Antonio Jordán-Pla (eds.), *Chromatin Immunoprecipitation: Methods and Protocols*,
Methods in Molecular Biology, vol. 1689, DOI 10.1007/978-1-4939-7380-4_7, © Springer Science+Business Media LLC 2018

[16]. Non-histone proteins may also protect against MNase digestion, but since most of these proteins are not forming the same tight contact with DNA, the assumption is that they do not protect against MNase digestion to the same extent as nucleosomes do. Nevertheless, histone-ChIP-seq should be performed to corroborate the nucleosome position [2, 3, 12], and labile nucleosomes have been detected at transcription start sites and termination sites in *Drosophila* cells [2]. One recent study suggests that labile nucleosomes mainly occupy AT-rich regions, for instance at transcription termination sites [3]. Mieczkowsk et al. [9] have employed different MNase concentrations to map nucleosomes with different sensitivities. This analysis, MACC (MNase accessibility), measures accessibility at different genomic regions and, when correlated with chromatin regions marked with specific histone modifications, transcription factors and chromatin factor, it shows that accessibility is higher in regions carrying active histone marks, such as H3K4me3 and H3k27Ac. In addition, limited MNase digestion generates DNA fragments of different lengths, and combined with ChIP seq, Lorzadeh et al. [6] correlated the longer DNA fragments with nucleosomes that carried the modification H3K27me3 at CpG islands. Nucleosomes carrying H3K4me3 gave instead fragments corresponding to single nucleosomes [6], suggesting that regions with H3K4me3 nucleosomes are less condensed. These studies provide us with tools to assess chromatin accessibility and relate the accessibility to different chromatin states.

Sites that are organized in open chromatin, usually found at regulatory sites such as promoters and enhancers, displaying an open chromatin structure, have been identified using DNAse I digestions [1, 4]. These DNAse I hypersensitive sites have been regarded as nucleosome-free regions. The MACC metric demonstrates that these sites are highly accessible, but highly sensitive nucleosomes are present [9]. This discrepancy can be explained by the difference in digestion activity; unlike MNase, DNAse I cuts DNA also inside nucleosomes, displaying distinct cleavage patterns depending on how the DNA interacts with the nucleosome [7]. Recently, DNAse I digestion in combination with sequencing has been used to map exposure of transcription factor binding sites in nucleosomes, giving fine-tuned maps of the rotational position of nucleosomes in chromatin regions [14, 15].

To assess the chromatin state at specific sites and compare those under different cellular conditions, MNase followed by qPCR is the appropriate way to analyze instead of sequencing. This type of analysis has been conducted to assess the nucleosome positions at the heat shock genes before and after activation by using the high concentration of MNase [11]. The accessibility of regions can be determined by limited MNase digestion, and this approach has been useful to assess changes in response to different treatments. In particular, non-annotated or repeated regions may be characterized in this way, and we have successfully mapped the rRNA promoter as well as the 5S rRNA loci by MNase digestion and primer walking [13, 15].

2 Materials

Prepare all solutions with pure water, distilled or MilliQ, and the highest grade of chemical purity.

1. Formaldehyde solution: 11% formaldehyde in 100 mM NaCl, 1 mM EDTA, 50 mM Hepes pH 7.6 (*see* **Note 1**).
2. Glycine solution: 2.5 M Glycine in distilled water.
3. PBS: 8 g of NaCl, 0.2 g of KCl, 1.44 g of Na_2HPO_4, 0.24 g of KH_2PO_4 is dissolved in 800 ml distilled water. Adjust the pH to 7.4 with HCl and add distilled water to a total volume of 1 l.
4. ChIP buffer 1: 50 mM HEPES at pH 7.6, 140 mM NaCl, 1 mM EDTA, 10% glycerol, 0.5% Nonidet P-40, 0.25% Triton X-100, in distilled water.
5. ChIP buffer 2: 10 mM Tris–Cl at pH 8, 200 mM NaCl, 1 mM EDTA, 0.5 mM EGTA, in distilled water.
6. 10 mM Tris–Cl, pH 7.5.
7. Micrococcal Nuclease (New England Biolabs, 2,000,000 gels units/ml), store at −20 °C.
8. MNase Digestion Buffer: 10 mM Tris-HCl at pH 7.5, 50 mM NaCl, 5 mM $MgCl_2$, 1 mM $CaCl_2$, 0.075% Nonidet P-40, in distilled water.
9. MNase Stop Buffer: 5% SDS, 50 mM EDTA, in distilled water.
10. Proteinase K solution: 20 mg/ml in distilled water.
11. PCI: Phenol:Chloroform:Isoamyl Alcohol (25:24:1).
12. Ethanol (95%).
13. 3 M Sodium Acetate, pH 5.2.
14. Glycogen 20 mg/ml.
15. For PCR analyses, appropriate primer pairs and SYBR Green analysis reagents.

3 Methods

3.1 Preparation of Chromatin

See **Note 2**.

We have successfully used the following protocol for studies of rDNA chromatin regulation [15]. We usually start with approximately 100,000–200,000 cells in 2 ml cell culture medium per treatment.

1. Treat the cells with 1% formaldehyde (final concentration) for 10 min at room temperature directly in the petri dish (*see* **Note 3**).
2. Stop the crosslinking by adding glycine solution to a final concentration of 125 mM and incubate for at least 10 min.
3. Wash the cells once in 2 ml PBS.
4. Lyse the cells in 2 ml ChIP buffer 1 at 4 °C (*see* **Note 4**).
5. Centrifuge the lysate at 500 × *g* for 5 min at 4 °C.
6. Wash the chromatin fraction (pellet) with 2 ml ChIP buffer 2.
7. Wash the pellet again with 1 ml 10 mM Tris–Cl at pH 7.5.
8. Dissolve the chromatin pellet in 0.3 ml MNase Digestion Buffer containing divalent ions.

3.2 MNase Digestion

1. Sonicate the chromatin fraction slightly (*see* **Note 5**).
2. Adjust the chromatin concentration of all samples (*see* **Note 6**).
3. Titrate the samples with MNase. Use 4 samples/treatment, starting with 0 U MNase and adding MNase in steps by 3 or 4 (for instance 0, 0.5 μl, 2 μl, and 8 μl MNase). A suitable reaction volume is 50–100 μl containing chromatin from 20,000 to 30,000 cells.
4. Incubate the samples with the different amounts of MNase for 15–30 min at 37 °C.
5. The MNase reaction is stopped by adding 5 μl MNase Stop solution. In addition, add 0.5 μl proteinase K to digest chromatin proteins during reversal of the cross-link.
6. The cross-link is reversed by heating the samples at least 5 h at 65 °C.
7. Isolate DNA fragments by PCI extraction, adding the same volume of PCI as in the samples. Precipitate the DNA with 2.5 volumes of 95% ethanol and 10 μl 3 M Sodium Acetate, for 1 h at −80 °C or overnight at −20 °C (*see* **Note 7**).
8. Dissolve each DNA sample in 100 μl distilled water.
9. The samples are then analyzed by quantitative real-time PCR (qPCR) (*see* **Note 8**). For high-resolution qPCR, design primers to produce 100 base-pair long products with 50 base pairs overlap covering the region of interest. *See* **Note 9** for interpretation of the qPCR analysis.
10. The degree of accessibility of different DNA sequences can be expressed in several ways:
    (a) The ratio between two titration results giving a lower value when more digested, i.e., more accessible DNA [11].
    (b) As % of digestion with 0 U MNase titration as baseline [4].

4 Notes

1. Prepare the formaldehyde solution using 37% formaldehyde, molecular biology grade.
2. Crosslinked chromatin is prepared as for ChIP-standard protocol (*see* for example reference [17]), but it is important to change to a buffer that is adequate for MNase digestion (a buffer with divalent ions). It is also possible to MNase digest permeabilized cells and then purify the fragmented DNA. For yeast, use a zymolyase treatment as in [12] or 0.01% L-α-lysophosphatidylcholine as in [15].
3. Add either 0.055 ml of 37% formaldehyde solution directly or 0.22 ml of the 11% formaldehyde solution for 2 ml cell culture medium (100,000–200,000 cells).
4. Adherent cells are scraped off and then resuspended in ChIP buffer 1, whereas cells in solution are directly resuspended in ChIP buffer 1.
5. We use a Bioruptor to sonicate the chromatin. When using 200,000 cells as starting material, optimal sonication is obtained with the 2-min program with 30 s sonication, 30 s rest. Increase the time of sonication slightly for high concentrations of chromatin.
6. It is important to start with approximately the same amount of chromatin in all samples. The chromatin concentration can be assessed by Bradford or by absorbance at 260 nm, and then adjusted to similar concentrations with MNase I reaction buffer. Suitable chromatin concentrations should give PCR cycles in the range of 15–30 when titrated. The starting concentration used depends on the nature of the genome regions; if repetitive regions are investigated, less chromatin is needed compared to single locus regions.
7. If the DNA concentration is low, use 5 μg of glycogen as the carrier to improve the yield of recovery.
8. DNA extracted from MNase-digested chromatin can also be sequenced (for example using Illumina parallel sequencing) [*see* for example reference 18 and Chapter 8].
9. The 0 U MNase titration and the higher amount of MNase should give different ct-values for all primer pairs used in the region analyzed and, for most samples, the two intermediate concentrations should be in-between the 0 U MNase and the highest MNase concentration used. It is important to analyze a few control regions with different accessibilities to control the performance of the experiment. If this is not possible because of the titration conditions, then primer pairs that cover the whole region, preferentially giving overlapping products,

should be used. Condensed DNA should be more enriched in the 0 U MNase titration than in the high MNase samples, and the more accessible DNA should be more represented in the MNase titration with the highest concentration.

References

1. Boyle AP, Davis S, Shulha HP, Meltzer P, Margulies EH, Weng Z, Furey TS, Crawford GE (2008) High-resolution mapping and characterization of open chromatin across the genome. Cell 132:311–322
2. Chereji RV, Kan TW, Grudniewska MK, Romashchenko AV, Berezikov E, Zhimulev IF, Guryev V, Morozov AV, Moshkin YM (2016) Genome-wide profiling of nucleosome sensitivity and chromatin accessibility in Drosophila melanogaster. Nucleic Acids Res 44 (3):1036–1051. doi:10.1093/nar/gkv978
3. Chereji RV, Ocampo J, Clark DJ (2017) MNase-sensitive complex in yeast: nucleosomes and non-histone barriers. Mol Cell 65 (3):565–577.e3. doi:10.1016/j.molcel.2016.12.009
4. Crawford GE, Davis S, Scacheri PC, Renaud G, Halawi MJ, Erdos MR, Green R, Meltzer PS, Wolfsberg TG, Collins FS et al (2006) DNase-chip: a highresolution method to identify DNase I hypersensitive sites using tiled microarrays. Nat Methods 3:503–509
5. Field Y, Kaplan N, Fondufe-Mittendorf Y, Moore IK, Sharon E, Lubling Y, Widom J, Segal E (2008) Distinct modes of regulation by chromatin encoded through nucleosome positioning signals. PLoS Comput Biol 4: e1000216
6. Lorzadeh A, Bilenky M, Hammond C, Knapp DJ, Li L, Miller PH, Carles A, Heravi-Moussavi A, Gakkhar S, Moksa M, Eaves CJ, Hirst M (2016) Nucleosome density ChIP-seq identifies distinct chromatin modification signatures associated with MNase accessibility. Cell Rep 17(8):2112–2124. doi:10.1016/j.celrep.2016.10.055
7. Lutter LC (1979) Precise location of DNase I cutting sites in the nucleosome core determined by high resolution gel electrophoresis. Nucleic Acids Res 6:41–56
8. Mavrich TN, Ioshikhes IP, Venters BJ, Jiang C, Tomsho LP, Qi J, Schuster SC, Albert I, Pugh BF (2008) A barrier nucleosome model for statistical positioning of nucleosomes throughout the yeast genome. Genome Res 18:1073–1083
9. Mieczkowski J, Cook A, Bowman SK, Mueller B, Alver BH, Kundu S, Deaton AM, Urban JA, Larschan E, Park PJ, Kingston RE, Tolstorukov MY (2016) MNase titration reveals differences between nucleosome occupancy and chromatin accessibility. Nat Commun 7:11485. doi:10.1038/ncomms11485
10. Noll M, Kornberg RD (1977) Action of micrococcal nuclease on chromatin and the location of histone H1. J Mol Biol 109:393–404
11. Petesch SJ, Lis JT (2008) Rapid, transcription-independent loss of nucleosomes over a large chromatin domain at Hsp70 loci. Cell 134 (1):74–84. doi:10.1016/j.cell.2008.05.029
12. Rodriguez J, McKnight JN, Tsukiyama T (2014) Genome-wide analysis of nucleosome positions, occupancy, and accessibility in yeast: nucleosome mapping, high-resolution histone ChIP, and NCAM. Curr Protoc Mol Biol 108:21.28.1–21.2816. doi:10.1002/0471142727.mb2128s108
13. Sadeghifar F, Böhm S, Vintermist A, Östlund Farrants AK (2015) The B-WICH chromatin-remodelling complex regulates RNA polymerase III transcription by promoting Max-dependent c-Myc binding. Nucleic Acids Res 43 (9):4477–4490. doi:10.1093/nar/gkv312
14. Vierstra J, Wang H, John S, Sandstrom R, Stamatoyannopoulos JA (2014) Coupling transcription factor occupancy to nucleosome architecture with DNase-FLASH. Nat Methods 11:66–72
15. Vintermist A, Böhm S, Sadeghifar F, Louvet E, Mansén A, Percipalle P, Ostlund Farrants AK (2011) The chromatin remodelling complex B-WICH changes the chromatin structure and recruits histone acetyl-transferases to active rRNA genes. PLoS One 6(4):e19184. doi:10.1371/journal.pone.0019184
16. Weiner A, Hughes A, Yassour M, Rando OJ, Friedman N (2010) High-resolution nucleosome mapping reveals transcription-dependent promoter packaging. Genome Res 20:90–100
17. Weinmann AS, Farnham PJ (2002) Identification of unknown target genes of human transcription factors using chromatin immunoprecipitation. Methods 26(1):37–47
18. Zhong J, Luo K, Winter PS, Crawford GE, Iversen ES, Hartemink AJ (2016) Mapping nucleosome positions using DNase-seq. Genome Res 26(3):351–364. doi:10.1101/gr.195602.115

Chapter 8

Characterization of the Nucleosome Landscape by Micrococcal Nuclease-Sequencing (MNase-seq)

Wieteke Anna Maria Hoeijmakers and Richárd Bártfai

Abstract

MNase-seq allows the genome-wide examination of the nucleosome landscape by determination of nucleosome positioning and occupancy. Typically, native or formaldehyde fixed chromatin is subjected to digestion by micrococcal nuclease (MNase), which degrades linker DNA and yields mainly mononucleosomes. The resulting material can be processed directly or can be subjected to an optional chromatin immunoprecipitation step (MNase-ChIP-seq). De-crosslinked and purified DNA is then subjected to next-generation sequencing. The protocol presented here has been tailored for the analysis of nucleosome landscape in the malaria parasite, *Plasmodium falciparum*, but most steps are directly applicable to other cell types. We also discuss general considerations for experimental design and computational analysis, which are crucial for accurate investigation of the nucleosome landscape.

Key words Micrococcal nuclease (MNase), MNase-seq, MNase-ChIP-seq, Nucleosome landscape, Nucleosome positioning, Nucleosome occupancy, AT-rich DNA, *Plasmodium falciparum*

1 Introduction

In all eukaryotes, the genetic material is packaged into chromatin. The basic unit of the chromatin is the nucleosome, consisting of two copies of each histoneMalaria type (H2A, H2B, H3, and H4). However, these nucleosomes do not evenly inhabit the genome, which has profound effects on the chromatin structure and accessibility of the underlying DNA sequences. Transcriptionally silent, heterochromatic regions are constantly occupied by densely packed nucleosomes, while active promoter regions are more accessible and characterized by the combination of a nucleosome-depleted region and a downstream array of well-positioned nucleosomes. The basic layout of the nucleosome landscape is dictated by the sequence preference of the histone octamer (that defines the position of certain nucleosomes) and the interaction between histone octamers (that governs nucleosome spacing). Chromatin-modifying enzymes actively adapt this landscape, which makes the

Neus Visa and Antonio Jordán-Pla (eds.), *Chromatin Immunoprecipitation: Methods and Protocols*, Methods in Molecular Biology, vol. 1689, DOI 10.1007/978-1-4939-7380-4_8, © Springer Science+Business Media LLC 2018

chromatin a highly dynamic structure. The resulting nucleosome landscape is typical to each cell and highly informative in deciphering chromatin function. For comprehensive reviews on nucleosome landscape, *see* [1, 2].

Micrococcal nuclease (MNase) has been isolated from *Staphylococcus aureus* in the 1960s and was among the first enzymes used to study the chromatin structure [3]. MNase is an endo/exonuclease, which preferentially digests single-stranded DNA, but can nick double-stranded DNA and digest RNA [4]. Importantly, it has a marked preference for linker DNA over DNA wrapped around the histone octamer or occupied by other proteins, making it an ideal enzyme to isolate nucleosomal DNA. MNase was first used to separate hetero- and eu-chromatin or study chromatin structure of individual (trans)genes, but with the rise of the genomic technologies it quickly became the workhorse to study the nucleosome landscape. MNase-seq utilizes next-generation sequencing technology to obtain millions of short sequence reads from the ends of nucleosomal DNA fragments, which can be used to decipher the position and occupancy of nucleosomes on a genome-wide scale [5, 6].

While several other methods have been developed to study the nucleosome landscape (MPE-seq [7]; ChIP-exo [8]; "hydroxy-radical-seq" [9]; "CAD-seq" [10]), MNase-seq remains to be the most used approach for this purpose. Here, we shortly discuss important considerations for the experimental design, caution for potential artifacts and describe a detailed protocol, which we used to decipher the nucleosome landscape of the malaria parasite, *Plasmodium falciparum* [11].

1.1 Experimental Considerations

While the MNase-seq procedure is reasonably straightforward, there are some key steps, which can markedly influence the outcome and interpretation of the results. Here, we highlight some points of consideration. A more detailed description of potential artifacts introduced in each of these steps can be found in the supplementary results and discussion of [11].

1.1.1 Formaldehyde Crosslinking and Chromatin Immunoprecipitation

In most MNase-seq protocols formaldehyde crosslinking is used to "fix" protein–DNA interactions and avoid movement of nucleosomes during MNase digestion. While in general this is advisable it is worthwhile to keep in mind that formaldehyde crosslinking preferentially crosslinks GC-rich sequences [12] and can crosslink non-histone proteins to the DNA as well. To specifically select for nucleosomal DNA fragments chromatin immunoprecipitation using an antibody against H3 or H4 is being used (MNase-ChIP-seq). This approach, on the other hand, can artificially alter the recovery of DNA from nucleosomes where the epitope is modified (e.g., carry posttranslational modification) or inaccessible (e.g., masked by other proteins) or even further lower the presence of AT-rich DNA fragments which are washed away due to their lower crosslinking efficiency (see above).

1.1.2 Titration of Digestion Conditions

Perhaps the most delicate part of the MNase-seq protocol is to define the digestion condition that yields mainly mononucleosomal fragments with a small amount of dinucleosomal DNA present. This has to be titrated for nearly all samples individually by varying the time of digestion or enzyme concentration applied. Sequencing of multiple digests from the same chromatin sample, as an alternative, can provide a more robust measure of chromatin accessibility [13]. Importantly, differences in digestion conditions for the individual samples can confound inter-sample comparison of the MNase-seq data. While this can be improved to some extent by standardized sample preparation [14], it is advisable to check and match the "extent" of digestion between samples. In addition, it is recommended to mix small amounts of nuclei from an unrelated species (e.g., yeast) to each sample in a similar proportion that can be used later on to correct for differences in digestion condition (spike-in control). This is particularly relevant if global differences in nucleosome occupancy are expected between the samples (e.g., upon deletion of a histone chaperone).

1.1.3 Size Selection

In early protocols, mononucleosomal-sized fragments have been selected during the preparation of sequencing libraries. This, however, can introduce unwanted technical variation by cutting out slightly different size ranges from the gel or when comparing samples with slightly different "extents" of digestion and leads to the removal of informative sub-nucleosomal or di-nucleosomal fragments. Instead, we advise omitting the size-selection step and use of paired-end sequencing to enable in silico size selection if necessary (*see* also later).

1.1.4 MNase Digestion Bias

Arguably, the biggest drawback of using MNase to decipher the nucleosome landscape is the intrinsic preference of this enzyme toward AT-rich sequences [15]. This can lead for example to over-digestion and artificial depletion of AT-rich nucleosomal DNA [11] and hence overestimation of the repulsive effect of AT-rich sequences for nucleosome formation [9]. Different approaches have been proposed to deal with these biases using computational methods [16] or using different reagents to fragment the chromatin [7, 9, 10]. The protocol we present here makes use of a combined MNase and exonuclease III treatment to alleviate some of the biases associated with the use of MNase alone [11, 17].

1.1.5 Biases in Next-Generation Sequencing

Biases during library preparation for next-generation sequencing and mapping of sequence reads can also negatively impact proper interpretation of the MNase-seq data [18]. In particular, the PCR-step common to most library preparation methods can bias against AT- and GC-rich sequences and against longer (e.g., di- or tri-nucleosomal) DNA fragments. Therefore, it is advisable to use PCR-free library preparation methods if possible [19, 20] or minimize the number of PCR cycles, increase the elongation time and use polymerases with reduced bias [11]. In addition, sequencing of fragmented genomic DNA (e.g., sonicated, MNase-digested, or both) can be used to correct for most biases related to library preparation and reduced mappability of sequencing reads [11] and is therefore recommended as a control in any MNase-seq procedures.

PCR can also lead to stochastic over-amplification of individual DNA fragments, which could for example lead to artificial "highly positioned" nucleosomes. While in silico removal of duplicate reads (i.e., sequence reads that start and end at the same position) is a commonly used approach to deal with this problem, in case of MNase-seq it can potentially lower the signal from very highly positioned nucleosomes. Therefore, the use of unique molecular identifiers (UMIs, [21]) to uniquely label each DNA fragment with a random sequence tag prior to PCR amplification is the most appropriate approach to deal with this problem.

1.2 Data Analysis

The computational analysis of MNase-seq data deserves a discussion that is beyond the volume of this chapter. Therefore, here we only list some typical steps in data analysis and like to refer our readers to excellent reviews on the subject [22, 23]. Sequence reads obtained from MNase-seq libraries are first mapped against the reference genome to generate a coverage plot [22]. As mentioned above, we recommend the use of paired-end sequencing, which provides more exact information about the coupling between digestion events and hence the fragment length. As a consequence, paired-end sequencing provides more accurate information on the nucleosome landscape and also enables in silico selection of fragments of various sizes (fragments smaller than 80 bp for example can be used to identify transcription factor binding or chromatin remodeling events [24] while dinucleosomal fragment sizes can be informative to confer linker length). Notably, appropriate use of the gDNA-seq control will require in silico correction of the fragment-length distribution of this control library to match the MNase-seq libraries as in [11]. The local distribution of MNase-seq fragment end- or mid-points can then be used to identify positioned nucleosomes. Comparison between MNase-seq datasets obtained from for example different stages of development can help to pinpoint dynamic changes in the nucleosome landscape [11]. Heat-maps or "averaged" nucleosome profiles can be used to investigate the nucleosome behavior surrounding specific genomic elements (e.g., transcription start sites, transcription factor binding sites, splice sites, or transcription end sites) for example in relation to gene expression. Nonetheless, often tailored analysis will need to be devised to answer specific research questions.

2 Materials

2.1 General Reagents

1. Nuclease-free water.
2. 100% glycerol.
3. QIAquick Spin Columns and reagents (Qiagen), or equivalent.
4. Qubit dsDNA HS kit (Invitrogen), or equivalent.

2.2 Formaldehyde Cross-Linking and Nuclei Isolation from P. falciparum Asexual Blood-Stage Parasites

1. *P. falciparum* asexual RBC stage culture in 5% hematocrit (*see* **Note 1**).
2. Plasmodipur filter (EuroProxima) (optional: 70% ethanol to sterilize the filter) (*see* **Notes 2** and **3**).
3. 10, 20, or 50 ml Syringe.
4. 37% formaldehyde (HCHO) solution (*see* **Note 4**).
5. 1.25 M Glycine solution.
6. Phosphate Buffered Saline (PBS) (chilled to 4 °C).
7. 10% saponin solution (*see* **Note 5**).
8. Cell Lysis Buffer (CLB): 10 mM Tris–HCl pH 8.0, 3 mM $MgCl_2$, 0.2% IGEPAL CA-630/Nonidet P-40, protease inhibitor cocktail (Complete, EDTA-free protease inhibitor cocktail, Roche, *see* **Note 6**). Chill to 4 °C before use.
9. Cell Lysis Buffer Sucrose (CLBS): CLB with 0.25 M sucrose. Chill to 4 °C before use.

2.3 Chromatin Digestion

1. MNase Storage Buffer: 20 mM HEPES pH 7.6, 50 mM NaCl, 50% glycerol.
2. Micrococcal nuclease (MNase; Worthington Biochemical Cooperation): reconstitute lyophilized powder to 20 units/μl stock in MNase Storage Buffer (*see* **Note 7**, store at −80 °C). Dilute to 0.5 units/μl in MNase Storage Buffer and aliquot to single-use aliquots, store at −80 °C.
3. Exonuclease III (100 units/μl; New England Biolabs).
4. Digestion Buffer (DB): 50 mM Tris–HCl pH 7.4, 4 mM $MgCl_2$, 1 mM $CaCl_2$, 0.075% IGEPAL CA-630/Nonidet P-40, 1 mM DTT (*see* **Note 8**), protease inhibitor cocktail (Complete, EDTA-free protease inhibitor cocktail, Roche, *see* **Note 6**). Chill to 4 °C before use.
5. Quenching Solution (QS): 2% Triton X-100, 0.6% SDS, 300 mM NaCl, 6 mM EDTA. Chill to 4 °C before use.

6. Sonication Buffer (SB): 25 mM Tris pH 7.4, 1% Triton X-100, 0.3% SDS, 150 mM NaCl, 3 mM EDTA, 2 mM $MgCl_2$, 0.5 mM $CaCl_2$, protease inhibitor cocktail (Complete, EDTA-free protease inhibitor cocktail, Roche, *see* **Note 6**). Chill to 4 °C before use.
7. 10% sodium dodecyl sulfate (SDS) solution (*see* **Note 9**).
8. 1 M sodium bicarbonate ($NaHCO_3$) solution.
9. 5 M sodium chloride (NaCl) solution.

2.4 Chromatin Immunoprecipitation (ChIP): Optional

1. ProtA Dynabeads (Life Technologies).
2. ProtG Dynabeads (Life Technologies).
3. Magnetic rack for Eppendorf tubes (Life Technologies DynaMag™-2, or equivalent).
4. Rotating wheel, at 4 °C.
5. Rotating wheel, at RT.
6. ChIP-grade α-histone-core antibody (*see* **Note 10**).
7. Beads Wash Buffer (BWB): 20 mM Tris–HCl pH 8.0, 2 mM EDTA, 1% Triton-X100, 0.1% SDS, 150 mM NaCl, protease inhibitor cocktail (Complete, EDTA-free protease inhibitor cocktail, Roche, *see* **Note 6**), 0.1% BSA (*see* **Note 11**). Chill to 4 °C before use.
8. ChIP Dilution Buffer (CDB): 20 mM Tris–HCl pH 8.0, 2 mM EDTA, 1% Triton X-100, 150 mM NaCl, protease inhibitor cocktail (Complete, EDTA-free protease inhibitor cocktail, Roche, *see* **Note 6**). Chill to 4 °C before use.
9. ChIP Wash Buffer 1 (CWB1): 20 mM Tris–HCl pH 8.0, 2 mM EDTA, 1% Triton X-100, 0.1% SDS, 150 mM NaCl. Chill to 4 °C before use.
10. ChIP Wash Buffer 2 (CWB2): 20 mM Tris pH 8.0, 2 mM EDTA, 1% Triton X-100, 0.1% SDS, 500 mM NaCl. Chill to 4 °C before use.
11. ChIP Wash Buffer 3 (CWB3): 10 mM Tris-HCl pH 8.0, 1 mM EDTA. Chill to 4 °C before use.
12. ChIP Elution Buffer (CEB): 1% SDS, 0.1 M $NaHCO_3$. Use at RT.
13. 10% Triton X-100.

2.5 Sequencing Library Preparation

For reagents and equipment used during library preparation, we refer to the list provided in [11, 25].

2.6 Equipment

1. Ice-buckets and ice.
2. Shaking, temperature-controlled incubator, at 37 °C.
3. Temperature-controlled centrifuge, at 4 °C.

4. Temperature-controlled benchtop centrifuge, at 4 °C.
5. Benchtop centrifuge, at RT.
6. Temperature-controlled water bath, at 37 °C.
7. Thermometer.
8. Shaking, temperature-controlled thermoblock, at 45 °C.
9. Qubit fluorimeter (Invitrogen), or equivalent.
10. Liquid nitrogen (for snap-freezing).
11. Sonicator (Bioruptor™ Next gen Sonicator (Diagenode)), or equivalent.
12. Glass homogenizer (7 ml) with type B pestle (Kontes).
13. Agarose gel electrophoresis system and imager, or Experion (BioRad), or Bioanalyzer (Agilent), or equivalent, and reagents.
14. DNase/RNase-free, Polyethylene Terephthalate (PET), 15 ml and/or 50 ml centrifugation tubes (*see* **Note 12**).
15. DNase/RNase-free Reaction Tubes (Eppendorf).
16. DNase/RNase-free filter tips.

3 Methods

Carry out all procedures on ice, unless stated otherwise. Use clean solutions, workspace, and filter tips to avoid contamination (*see* **Note 13**).

3.1 Formaldehyde Cross-Linking and Nuclear Isolation from P. falciparum *Asexual Blood Stage Cultures (See Notes 14 and 15)*

1. Mount the Plasmodipur filter on a syringe (*see* **Notes 2**, **3**, and **16**).
2. Take the parasite culture from 37 °C, pass through the Plasmodipur filter and collect in a 15 or 50 ml PET centrifugation tube (*see* **Note 12**). Immediately add 37% formaldehyde solution at a concentration of 1% formaldehyde. For *P. falciparum* asexual stages, minimally use the number of parasites specified in Table 1.

Table 1
The appropriate amount of blood-stage, asexual *P. falciparum* parasites minimally required to obtain two aliquots of about 8 μl nuclei for MNase-seq

Plasmodium falciparum stage	Approximate amount of parasites	Approximate culture volume, ml
Ring	1.7×10^9 parasites	30
Trophozoite	7.2×10^8 parasites	13
Schizont	1.8×10^8 parasites	3.5

One aliquot can be used for the "test" digestion, the other aliquot can be used for the "real" digestion. For the estimation of the required culture volume, parasites are cultured to ~10% parasitemia in 5% hematocrit

3. Incubate for 15 min at 37 °C while shaking.
4. Quench the reaction by addition of 1.25 M glycine to reach a final concentration of 0.125 M glycine. Transfer the culture to ice.
5. Spin the fixed cells for 5 min, 600 × *g* at 4 °C.
6. Remove the supernatant and wash the pellet with 1 culture volume of chilled PBS by gentle resuspension. Spin for 5 min, 600 × *g* at 4 °C.
7. Remove the supernatant and gently resuspend the pellet in 1 culture volume of chilled PBS.
8. Add 10% saponin to obtain a final concentration of 0.05% to lyse the RBCs. Gently turn the tube upside down 6× to mix the solution and place back on ice.
9. Incubate for 1–5 min on ice until lysis is observed (*see* **Note 17**).
10. Spin the tube for 10 min, 2400 × *g* at 4 °C.
11. Remove the supernatant and wash the pellet twice with 1 culture volume chilled PBS. Resuspend the pellet well and spin for 10 min, 2400 × *g* at 4 °C at each wash step (*see* **Notes 18** and **19**).
12. After the final wash remove as much supernatant as possible to obtain clean isolated *P. falciparum* parasites (*see* **Note 20**).
13. Resuspend the isolated parasites in $^1/_3$ culture volume CLB.
14. Gently homogenize with a glass homogenizer on ice, using 10–15 strokes.
15. Prepare a 15 ml PET centrifugation tube containing $^1/_3$ culture volume CLBS.
16. Collect the nuclei in CLB (from **step 14**) from the homogenizer and carefully layer them over the CLBS solution (*see* **Note 21**).
17. Spin in a swing-out centrifuge for 10 min at 2400 × *g* at 4 °C.
18. Carefully remove all supernatant and resuspend the nuclei in 1–2 ml CLB supplemented with 25% glycerol. Aliquot to at least two equally sized aliquots of approximately 5–8 μl worth of nuclei each (*see* **Note 22**). Spin aliquots for 10 min at 2400 × *g* 4 °C, remove the supernatant as much as possible, snap-freeze and store at −80 °C (*see* **Note 23**).

3.2 Chromatin Digestion

3.2.1 "Test" Digestions (See **Notes 24** and **25**, and Fig. 1)

19. Heat the water bath to 37 °C and check the temperature with a thermometer.
20. Prepare five Eppendorf tubes containing 25 μl QS for each aliquot of frozen nuclei that will be used for test-digestion. Place them on ice.
21. Determine for each pellet the digestion time points that will be tested (*see* **Note 26**). From each aliquot five different reaction time points can be tested in the range of 3–15 min.
22. Remove the nuclei from −80 °C storage and thaw on ice.
23. Resuspend nuclei in 1 ml DB each and spin for 16 min at 2400 × *g* at 4 °C.

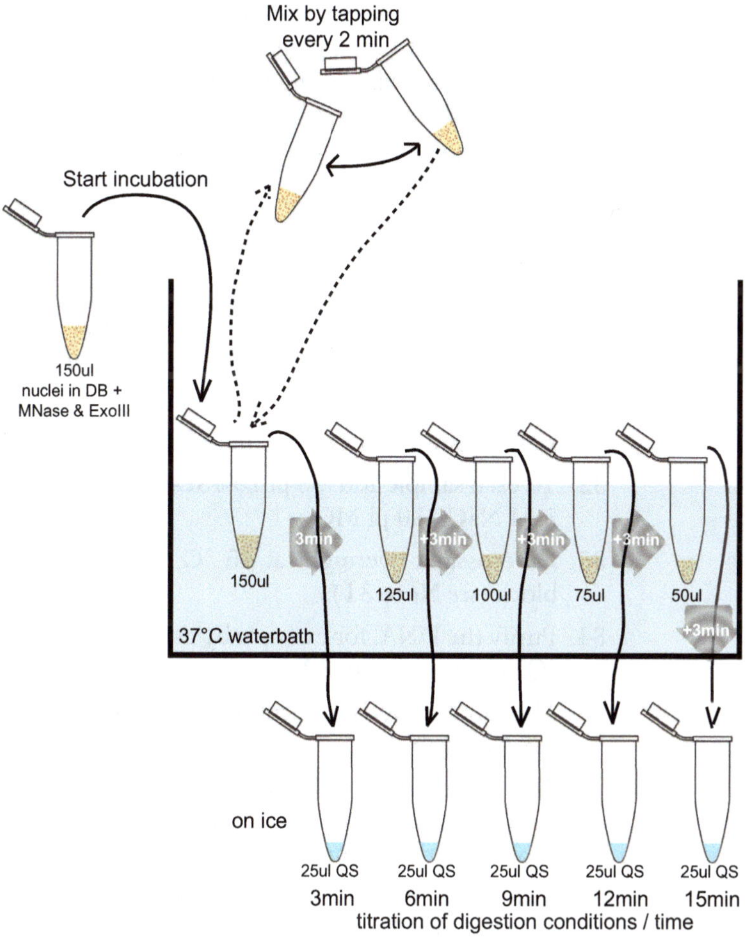

Fig. 1 Schematic representation of "test" digestion setup. After addition of MNase and exonuclease III enzymes, the nuclei in digestion buffer (DB) are transferred to a 37 °C water bath. During the 37 °C incubation, regular mixing is required to prevent settling of the nuclei at the bottom of the tube. From a single nuclei aliquot, five different digestion times can be tested by sequentially removing 25 μl from the 150 μl reaction tube at every time point and adding this to 25 μl ice-cold QS

24. Remove all supernatant and resuspend the nuclei in 150 μl DB.
25. Add to each 150 μl reaction 0.5 units MNase and 100 units Exonuclease III. Mix by tapping and place in water bath. Start your timer directly.
26. Keep the nuclei in suspension by briefly tapping the tube every 2 min (*see* **Note 27**) during the incubation.
27. Once the shortest digestion is completed, mix the nuclei by tapping and quickly remove 25 μl solution. Place the remaining digestion reaction back into the 37 °C water bath immediately and add the 25 μl test-digest to one of the Eppendorf tubes containing 25 μl QS and resuspend well. Leave the tube on ice until all digestions are over.
28. Once the next-shortest digestion is completed, again mix the nuclei in DB by tapping and quickly remove 25 μl solution. Place the remaining digestion reaction back into the 37 °C water bath immediately and add the 25 μl test-digest to one of the Eppendorf tubes containing 25 μl QS, resuspend well and put on ice.
29. Repeat **step 28** until all five test-digests per aliquot are in QS on ice (*see* **Note 28**).
30. Fill up the volume in each tube to 300 μl by addition of SB and briefly sonicate the samples to more effectively free the chromatin from the nuclear debris (*see* **Note 29**).
31. Spin samples for 10 min, 9600 × *g*, at 4 °C and transfer the supernatant to a new Eppendorf tube (*see* **Note 30**).
32. To each sample add: 35 μl 10% SDS, 50 μl 1 M $NaHCO_3$, 85 μl 5 M NaCl, 30 μl MQ.
33. De-crosslink overnight at 45 °C, while shaking in a thermoblock (*see* **Note 31**).
34. Purify the DNA, for example by using QIAquick Spin Columns (Qiagen) (*see* **Note 32**).
35. Measure the DNA concentration, for example by using dsDNA HS kit (Qubit).
36. Assess digestion rate by analyzing each sample on a 2% agarose gel (use 100-200 ng DNA/sample) or Experion (BioRad) or Bioanalyzer (Agilent) system along a 100 bp DNA ladder.
37. Determine the most optimal digestion condition for each sample as in Fig. 2 (*see* **Note 33**).

3.2.2 "Real" Digestion (See **Notes 34** and **35**)

38. Once the most optimal digestion condition for each sample has been determined, the final digestions can be performed.

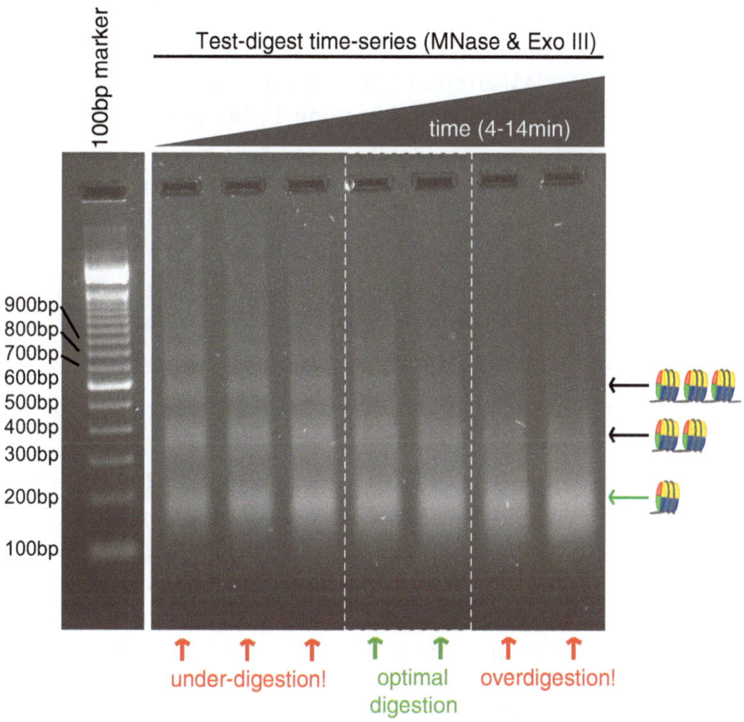

Fig. 2 2% agarose gel of a typical test digestion of seven tested time-points ranging from 4 to 14 min. Mononucleosomal, dinucleosomal, and trinucleosomal bands are *highlighted* and the two most optimal digestion times are *boxed*. Overdigestion is evident from loss-of-sharpness of and increased smearing below the mononucleosomal DNA band

39. Perform **steps 19** and **22–26** identical to what has been done in the test digestions. This time only a single digestion length for the entire 150 μl sample is used.
40. Once the digestion is completed, remove the tube from the 37 °C water bath and immediately add 150 μl QS. Resuspend well and place on ice (*see* **Note 28**).
41. Sonicate the samples as in **step 30**.
42. Spin samples for 10 min, 9600 × *g*, at 4 °C and transfer the supernatant to a new Eppendorf tube (*see* **Notes 30** and **36**).
43. Take 50 μl of chromatin for de-crosslinking and add: 18.5 μl 10% SDS, 20 μl 1 M $NaHCO_3$, 38.5 μl 5 M NaCl, 73 μl MQ. The remainder of the chromatin could be used for the optional ChIP procedure (*see* Subheading 3.3 and **Note 37**).
44. Perform **steps 33–35**.

45. The isolated DNA can be directly used to continue with Subheading 3.4 in case of MNase-seq without ChIP (called "MNase-seq"; for consideration on whether to include ChIP or not, *see* Subheading 1) (*see* **Note 38**).

3.3 Chromatin-Immunoprecipitation (ChIP): Optional (See Subheading 1 and Note 39)

46. Calculate the DNA concentration in each chromatin sample from the DNA concentration measurement of the de-crosslinked chromatin sample from **step 44**.
47. For each ChIP use 200 ng of chromatin (*see* **Notes 40** and **41**) and fill up with SB to 50 μl.
48. Add 50 μl CDB to each 50 μl chromatin sample in SB.
49. Add 1 μg antibody to each reaction and rotate at 4 °C overnight.
50. Prepare the beads as follows: Mix 10 μl ProtA beads solution with 10 μl ProtG beads solution for each ChIP reaction (*see* **Note 42**). Wash the beads twice with 1 ml BWB as follows: add buffer, invert tube 10×, spin 6 s to max 400 × *g*, place in a magnetic holder for 1 min and remove the liquid without touching the beads. After the last wash reconstitute the beads in CWB1 bringing the total volume back to 20 μl per ChIP reaction.
51. Add 20 μl of well-mixed ProtA/ProtG-bead slurry to each sample and rotate for 2 h at 4 °C.
52. Take samples from rotation wheel, spin 6 s to max 400 × *g*, place in a magnetic holder for 30 s and remove the supernatant.
53. Wash 1× using CWB1, 2× using CWB2, and 2× using CWB3. For each wash add 400 μl of appropriate wash buffer, rotate for 5 min at 4 °C, spin 6 s to max 400 × *g*, place in a magnetic holder for 30 s and remove wash buffer.
54. After the final wash, remove all liquid and bring the samples to RT. Then add 200 μl CEB and mix well. Rotate for 20 min at RT.
55. Briefly spin 6 s to max 400 × *g* at RT, place in a magnetic holder at RT and transfer the eluate to a new Eppendorf tube (keep at RT) (*see* **Note 43**).
56. To each 200 μl ChIP eluate add: 10 μl 10% SDS, 10 μl 1 M $NaHCO_3$, 60 μl 5 M NaCl, 10 μl 10% Triton X-100, 10 μl H_2O (*see* **Note 44**).
57. To each 100 μl input sample add: 27 μl 10% SDS, 30 μl 1 M $NaHCO_3$, 57 μl 5 M NaCl, 86 μl H_2O (*see* **Note 44**). Alternatively, the input DNA isolated at **step 45** can be used here.
58. Decrosslink, purify, and determine the DNA concentration as prescribed in **steps 33–35** (*see* **Notes 32** and **45**).

3.4 Sequencing Library Preparation (See Note 46)

59. Use 1–10 ng "MNase-seq", "MNase-ChIP-seq," and "gDNA control" DNA per library preparation reaction (*see* Subheading 1 for more details on the controls).
60. For library preparation follow the detailed LADS protocol provided in Hoeijmakers et al. [25]. Include the modifications prescribed in Kensche et al. [11] to allow inclusion of barcoded adapters and KAPA polymerase mediated library amplification and to prevent library size-selection (also *see* Subheading 1).

4 Notes

1. **CAUTION**: *P. falciparum* asexual intra-erythrocytic stages are infectious via blood contact, so special care should be taken when handling sharp objects.
2. Do not use suspensions of higher than 50% red blood cells (RBCs) and do not process more than 12.5 ml packed RBC per Plasmodipur filter to ensure efficient removal of white blood cells.
3. As an alternative for filtering the iRBC culture during the collection for MNase-Seq, human blood can be pre-filtered before addition to the *P. falciparum* culture. This option is preferred as it will minimize handling time upstream of cross-linking, but can best be performed in combination with a final Percoll gradient centrifugation synchronization step as in [11]. As Plasmodipur filters are not supplied sterile from the manufacturer, the filter can be sterilized by extensive spraying of 70% ethanol solution to the outside of the filter followed by mounting a syringe containing 20 ml 70% ethanol on the filter and passing the ethanol solution through. Subsequently, the filter needs to be washed with 2 × 50 ml sterile MQ and 20 ml medium before filtering the blood solution. All these steps need to be performed in a sterile-flow cabinet.
4. **CAUTION**: Formaldehyde is toxic by ingestion, skin contact, and inhalation. Therefore, appropriate protection (gloves, labcoat, safety glasses) and a well-ventilated working environment are required. It is advised to carefully read and follow the detailed safety instructions provided by the manufacturer.
5. **CAUTION**: Saponin is toxic upon oral or dermal contact of by inhalation. Use of a dusk mask, safety glasses, and gloves is recommended when weighing and dissolving saponin powder.
6. Roche Protease Inhibitor needs to be added fresh.
7. **CRITICAL**: MNase enzyme solution needs to be aliquoted into single-use aliquots. Avoid freeze–thaw cycles to retain optimal enzyme activity. As enzyme efficiency might vary for

each reconstituted batch, make sure to use the same batch enzyme for "test" and "real" digestions.

8. DTT needs to be added fresh.
9. **CAUTION**: SDS is toxic upon oral or dermal contact of by inhalation. Use of a dusk mask, safety glasses, and gloves is recommended and weighing and dissolving SDS powder needs to be performed in a fume cabinet with appropriate ventilation. For more information, carefully review the manufacturer's safety instructions.
10. **CRITICAL**: It is critical to use a ChIP-grade α-histone-core antibody that will precipitate all nucleosomes irrespective of nucleosome variants or post-translational modifications if the aim is to assess the overall nucleosome landscape, for example use α-H3core (Abcam Ab1791, lot GR88948-1) or α-H4core (Abcam Ab17036, lot GR8733-1).
11. BSA should be added fresh.
12. Make sure to use Polyethylene Terephthalate (PET) when collecting *P. falciparum*-infected red blood cells (RBCs). When using centrifugation tubes consisting of other types of plastic and/or coating, the (infected)RBCs might adhere to the sides of the tube.
13. **CRITICAL**: It is critical to work in a clean area, with clean buffers and pipets and while using filter tips to avoid contamination of DNA from other sources in the final sequencing libraries.
14. The protocol for formaldehyde cross-linking and nuclear isolation described here has been optimized for RBCs infected with *P. falciparum* asexual stages. For any other cell type or life cycle stage, a dedicated protocol should be used.
15. **TIMING**: Subheading 3.1 will take roughly 3 h to complete. It will take approximately 2 h from the startMalaria of the protocol to complete **step 12** (isolate *P. falciparum* parasites) which introduces a potential **PAUSE POINT** in this procedure.
16. **CRITICAL**: Removal of contaminating human white blood cells (WBCs) is required to prevent sequencing large amounts of human DNA, which would severely compromise cost-efficiency.
17. Upon lysis of the RBCs, the culture will turn from opaque to a translucent red color. Sometimes this change is difficult to observe. If no shift in transparency is observed continue with centrifugation after 5 min incubation on ice.
18. After saponin lysis and centrifugation, often RBCs are still observed in the pellet (visible as a (partial) bright red pellet).

These will generally lyse upon resuspension during the downstream PBS washes. If substantial amounts of RBCs are still part of the pellet after the second wash add a third wash before continuing.

19. After saponin lysis, ring stage parasites often form a whitish layer on top of the pellet. To prevent loss of parasites, take care to not remove too much supernatant on top of the pellet during the wash steps.
20. **PAUSE POINT**: Parasites could be stored at this stage. To store the isolated parasites, resuspend the pellet in 1 ml PBS supplemented with 25% glycerol, transfer to an Eppendorf tube, spin for 10 min at 2400 × *g* at 4 °C, remove the supernatant as much as possible, snap-freeze and store at −80 °C for prolonged periods of time. Upon continuation of the protocol, first wash the parasites once in PBS to remove the glycerol and then continue with **step 13**.
21. Pipet gently and avoid the final "blow-out" from a pipet boy to prevent mixing of the layers. It is important that two layers are formed (CLBS on the bottom, CLB with nuclei on the top) to ensure clean isolation of nuclei and prevent carry-over of contaminants (e.g., hemoglobin).
22. **CRITICAL**: It is very important to prepare at least 2—but preferably more—aliquots of equal amounts of isolated nuclei. One or more aliquots can be used to test and optimize the digestion time required to generate predominantly mononucleosomal chromatin fragments. The final aliquot(s) should be used for the "real" digestion using the most optimal digestion conditions as determined from the testing-aliquots (*see* Figs. 1 and 2).
23. **PAUSE POINT**: Nuclei can be stored at −80 °C for years.
24. **CRITICAL**: The test digestions are a critical part of the procedure to optimize the right digestion conditions for each batch of nuclei or batch of MNase. This step in the protocol will prevent over- or under-digestion of individual chromatin samples and ensure comparable digestion between samples within a set to allow comparison of those samples. Failure to control (equal) digestion efficiency can result in artifacts as outlined in Subheading 1 (also *see* Supplementary Results and Discussion in [11]). To ensure equal conditions between "test" and "real" digestions warrant that nuclear aliquots are of equal size, the MNase enzyme is aliquoted from the same batch, the same water bath is used.
25. **TIMING**: The "*Test" digestions* in Subheading 3.2 will take roughly 2–3 h "handling" time followed by a 20–24 h overnight incubation and another 2–4 h "handing" time.

26. The enzyme amounts and digestion times indicated in the protocol are examples based on our experience and will likely vary depending on the biological material and the enzyme source. Therefore, each researcher will need to determine this for his-/herself.
27. To prevent nuclei to settle at the bottom of the tube while incubating in the water bath, it is important to regularly mix the tubes by gentle tapping. Make sure to minimize the time that the tubes are out of the 37 °C water bath to allow efficient digestion. Alternatively, a shaking temperature-controlled heat block could be employed. However, in our experience using a heat block less consistent results are obtained, especially for longer digestion times – likely the consequence of increased evaporation to the non-heated lid – therefore the use of a water bath is recommended.
28. **PAUSE POINT**: Quenched digestion reactions can be stored at −20 °C for prolonged periods of time.
29. The settings of the sonicator will depend on the machine available. We made use of a Bioruptor™ Next gen Sonicator (Diagenode) for six cycles of 10 s ON/10 s OFF using setting LOW (reach level 4). For the Bioruptor sonicator it is advised to transfer the sample to special TPX polymethylpentene tubes (Diagenode 1.5 ml cat# C30010010-300).
30. **PAUSE POINT**: Soluble, digested chromatin can be stored at 4 °C overnight or at −20 °C for prolonged periods of time.
31. For de-crosslinking procedure, prolonged overnight incubation for 20–24 h is advised to enhance efficiency of de-crosslinking. De-crosslinking at 45 °C instead of 65 °C is advised to reduce sequence bias. For more details *see* Supplementary Fig. 3 in [25].
32. **PAUSE POINT**: Purified DNA can be stored at −20 °C for prolonged periods of time.
33. Digestion efficiency of each test digestion can be assessed using 2% agarose gels (always load a 100 bp ladder in parallel). The optimal digestion condition results in primarily mononucleosomes (~150 bp size), with little di- or even trinucleosomal bands present. However, it is important that heavy over-digestion of the mononucleosomal band – characterized by increased smearing to lower size range products – is prevented. For an example test digestion *see* Fig. 2.
34. It is best practice to handle samples that need to be directly compared on the same day, using the same reagents and conditions.

35. **TIMING**: The "*Real*" *digestion* in Subheading 3.2 will take roughly 2 h "handling" time followed by a 20–24 h overnight incubation and another 30 min "handing" time.
36. **CONTROL**: To check whether specific chromatin domains remain associated with insoluble nuclear structures (e.g., membranes), the nuclear debris resulting after **step 31** may be resuspended in 50 μl SB and de-crosslinked overnight as in **step 33** (the pellet can either be used directly for de-crosslinking or can be stored at −20 °C for prolonged periods). The DNA-content from this material can be quantified to estimate the chromatin fraction that remains associated with insoluble nuclear structures and (a subset of) these samples can be further processed into sequencing libraries to assess whether specific chromatin domains become depleted during the procedure (MNase-Pellet-Seq). For *P. falciparum*, we have found that centromeres are preferentially retained in the insoluble pellet fraction thereby cautioning strong statements on nucleosome occupancy from these regions.
37. **PAUSE POINT**: Store the remainder of the chromatin as back-up or for later use in MNase-ChIP reactions. Snap-freeze the chromatin and store at −80 °C for years. Make aliquots to avoid many freeze–thaw-cycles.
38. **QUALITY CONTROL**: Load an aliquot of the DNA on a 2% agarose gel (use 100–200 ng DNA/sample) or Experion (BioRad) or Bioanalyzer (Agilent) system to check digestion efficiency, but ensure that sufficient material is left for sequencing library preparation of "MNase-seq input" samples.
39. **TIMING**: Subheading 3.3 will take roughly 30 min "handling" time followed by a ~16 h overnight incubation, ~4 h "handing" time, another overnight incubation (20–24 h) and finally ~30 min "handing" time. A **PAUSE POINT** can be introduced after **step 55** (ChIP elution).
40. If the chromatin amount is insufficient to use 200 ng DNA per ChIP reaction less chromatin might be used. To make all reactions comparable also reduce the amount of chromatin added to each reaction from the other chromatin samples.
41. **CONTROL**: Keep at least 100 ng chromatin aside as input. Fill input sample up to 50 μl by addition of SB and then add 50 μl of CDB, store at 4 °C until de-crosslinking in **step 57**. Alternatively, use the DNA isolated in **step 45** as input sample.
42. **CONTROL**: For each chromatin sample, besides performing the ChIP reaction, always include a negative control reaction (with negative IgG antibody). The ChIP and control reactions can be assessed by ChIP-qPCR and clear enrichment over negative control should be evident. In our hands, recoveries of negative control samples range 0.1–0.2% recovery for any

chromatin region, whereas recoveries for a α-H3core (Abcam Ab1791, lot GR88948-1) or α-H4core (Abcam Ab17036, lot GR8733-1) were in the range of 40% or 3% recovery, respectively.

43. **PAUSE POINT**: Eluted chromatin can be stored at −20 °C for prolonged periods of time. High concentrations of SDS result in precipitation under low-temperature conditions, therefore ensure to thaw the sample at RT until all precipitated SDS is back in solution before continuing on the protocol.
44. Do not make a pre-mix of the "to-be-added" solutions as the high concentration of salt and detergent will cause precipitation in solution. Instead just add each component separately to each tube.
45. **QUALITY CONTROL**: Perform a qPCR on the ChIP-material to verify that the ChIP procedure worked properly and confirm that the α-histone ChIP yielded recoveries in the range 3–40% (depending on the α-histone-core antibody used) while the negative IgG control ChIP yielded low recoveries (<0.5%).
46. **TIMING**: Subheading 3.4 will take approximately 1 day if a KAPA PCR amplification protocol is employed [11] or 2 days if the LADS procedure is employed [25]. Sequencing will take 2–5 days depending on the cycle number, sequencing of single or both ends and the type of sequencer used.

Acknowledgments

The research leading to this protocol has received funding from The Netherlands Organization for Scientific Research (NWO-Vidi 864.11.007 to R.B.) and The National Institutes of Health (EuPathDB-Driving Biological Project subaward # 553539 to R. B.). We would like to acknowledge Dr. Kensche for valuable input and discussions concerning data analysis and Christa Toenhake for proofreading of the manuscript. Furthermore, we would like to thank our colleagues at the Department of Molecular Biology, the Department of Molecular Developmental Biology, and the Department of Medical Microbiology of Radboud University and St. Radboud UMC for support and advice.

References

1. Lieleg C, Krietenstein N, Walker M, Korber P (2015) Nucleosome positioning in yeasts: methods, maps, and mechanisms. Chromosoma 124:131–151
2. Hughes AL, Rando OJ (2014) Mechanisms underlying nucleosome positioning in vivo. Annu Rev Biophys 43:41–63
3. Clark RJ, Felsenfeld G (1971) Structure of chromatin. Nat New Biol 229:101–106

4. Heins JN, Suriano JR, Taniuchi H, Anfinsen CB (1967) Characterization of a nuclease produced by Staphylococcus aureus. J Biol Chem 242:1016–1020
5. Cui K, Zhao K (2012) Genome-wide approaches to determining nucleosome occupancy in metazoans using MNase-Seq. Methods Mol Biol 833:413–419
6. Platt JL, Kent NA, Harwood AJ, Kimmel AR (2013) Analysis of chromatin organization by deep sequencing technologies. Methods Mol Biol 983:173–183
7. Ishii H, Kadonaga JT, Ren B (2015) MPE-seq, a new method for the genome-wide analysis of chromatin structure. Proc Natl Acad Sci U S A 112:E3457–E3465
8. Rhee HS, Bataille AR, Zhang L, Pugh BF (2014) Subnucleosomal structures and nucleosome asymmetry across a genome. Cell 159:1377–1388
9. Brogaard K, Xi L, Wang JP, Widom J (2012) A map of nucleosome positions in yeast at base-pair resolution. Nature 486:496–501
10. Allan J, Fraser RM, Owen-Hughes T, Keszenman-Pereyra D (2012) Micrococcal nuclease does not substantially bias nucleosome mapping. J Mol Biol 417:152–164
11. Kensche PR, Hoeijmakers WA, Toenhake CG, Bras M, Chappell L, Berriman M, Bartfai R (2016) The nucleosome landscape of *Plasmodium falciparum* reveals chromatin architecture and dynamics of regulatory sequences. Nucleic Acids Res 44:2110–2124
12. Orlando V, Strutt H, Paro R (1997) Analysis of chromatin structure by in vivo formaldehyde cross-linking. Methods 11:205–214
13. Mieczkowski J, Cook A, Bowman SK, Mueller B, Alver BH, Kundu S, Deaton AM, Urban JA, Larschan E, Park PJ, Kingston RE, Tolstorukov MY (2016) MNase titration reveals differences between nucleosome occupancy and chromatin accessibility. Nat Commun 7:11485
14. Rizzo JM, Bard JE, Buck MJ (2012) Standardized collection of MNase-seq experiments enables unbiased dataset comparisons. BMC Mol Biol 13:15
15. Horz W, Altenburger W (1981) Sequence specific cleavage of DNA by micrococcal nuclease. Nucleic Acids Res 9:2643–2658
16. Kaplan N, Hughes TR, Lieb JD, Widom J, Segal E (2010) Contribution of histone sequence preferences to nucleosome organization: proposed definitions and methodology. Genome Biol 11:140
17. Nikitina T, Wang D, Gomberg M, Grigoryev SA, Zhurkin VB (2013) Combined micrococcal nuclease and exonuclease III digestion reveals precise positions of the nucleosome core/linker junctions: implications for high-resolution nucleosome mapping. J Mol Biol 425:1946–1960
18. Meyer CA, Liu XS (2014) Identifying and mitigating bias in next-generation sequencing methods for chromatin biology. Nat Rev Genet 15:709–721
19. Oyola SO, Otto TD, Gu Y, Maslen G, Manske M, Campino S, Turner DJ, Macinnis B, Kwiatkowski DP, Swerdlow HP, Quail MA (2012) Optimizing Illumina next-generation sequencing library preparation for extremely AT-biased genomes. BMC Genomics 13:1
20. Hoeijmakers WA, Bartfai R, Francoijs KJ, Stunnenberg HG (2010) Linear amplification for deep sequencing. Nat Protoc 6:1026–1036
21. Kivioja T, Vaharautio A, Karlsson K, Bonke M, Enge M, Linnarsson S, Taipale J (2012) Counting absolute numbers of molecules using unique molecular identifiers. Nat Methods 9:72–74
22. Quintales L, Vazquez E, Antequera F (2014) Comparative analysis of methods for genome-wide nucleosome cartography. Brief Bioinform 16:576–587
23. Teif VB (2015) Nucleosome positioning: resources and tools online. Brief Bioinform 17:745–757
24. Henikoff JG, Belsky JA, Krassovsky K, MacAlpine DM, Henikoff S (2011) Epigenome characterization at single base-pair resolution. Proc Natl Acad Sci U S A 108:18318–18323
25. Hoeijmakers WA, Bartfai R, Francoijs KJ, Stunnenberg HG (2011) Linear amplification for deep sequencing. Nat Protoc 6:1026–1036

Chapter 9

ChIP-re-ChIP: Co-occupancy Analysis by Sequential Chromatin Immunoprecipitation

Timothy V. Beischlag, Gratien G. Prefontaine, and Oliver Hankinson

Abstract

Chromatin immunoprecipitation (ChIP) exploits the specific interactions between DNA and DNA-associated proteins. It can be used to examine a wide range of experimental parameters. A number of proteins bound at the same genomic location can identify a multi-protein chromatin complex where several proteins work together to regulate gene transcription or chromatin configuration. In many instances, this can be achieved using sequential ChIP; or simply, ChIP-re-ChIP. Whether it is for the examination of specific transcriptional or epigenetic regulators, or for the identification of cistromes, the ability to perform a sequential ChIP adds a higher level of power and definition to these analyses. In this chapter, we describe a simple and reliable method for the sequential ChIP assay.

Key words Chromatin immunoprecipitation (ChIP), ChIP-re-ChIP, Promoter/regulatory region occupancy, Co-occupancy, Polymerase chain reaction (PCR), Chemical cross-linking, Protein–protein interactions

1 Introduction

Over the last two decades, chromatin immunoprecipitation (ChIP) has been most commonly used in combination with the polymerase chain reaction (PCR) to confirm the association of DNA-binding proteins with specific regions of chromatin such as DNA response element binding and promoter/enhancer occupancy. Generally, these associations correlate with transcriptional activation/repression [1, 2] and alterations in chromatin conformation [3, 4] or to demonstrate the dynamic nature of transcription factor-promoter occupancy [5, 6].

The sequential ChIP or ChIP-re-ChIP, where one chromatin-associated protein entity is precipitated followed by a second round of precipitation with an antibody to a second entity, is employed to demonstrate that both proteins are associated with the same region of chromatin. An important caveat is that these proteins only exist in that same region of chromatin. The sequential ChIP has often

Neus Visa and Antonio Jordán-Pla (eds.), *Chromatin Immunoprecipitation: Methods and Protocols*,
Methods in Molecular Biology, vol. 1689, DOI 10.1007/978-1-4939-7380-4_9, © Springer Science+Business Media LLC 2018

been misinterpreted to mean that both entities are associated directly with each other by weak hydroelectric forces or are part of the same complex. In the instance where direct protein–protein interactions are being examined and, because of the nature of ChIP, one can only use sequential ChIP as correlative evidence that must be backed up by other biochemical means demonstrating protein–-protein interactions or other types of protein association. Thus, the sequential ChIP may be more informative for associating protein occupancy with alterations in chromatin conformation. These alterations are often indirectly identified by immunoprecipitation with appropriate marker antibodies such as those raised against modified histones or DNA methyl-binding proteins. Again, in this instance the proteins being examined may not exist in the same complex but may bind at different sites within the approximately 500–1000 bp fragment generated by sonication.

Regardless of the intended purpose, it is ideal to demonstrate sequential precipitation of co-occupancy protein components in either order. Specifically, in the situation where an antibody directed to one component protein of a co-occupancy complex is used for the first ChIP, and a second antibody directed to the other protein component is used for the second ChIP, then the reverse combination of precipitations should also be demonstrated. This practice should be performed whenever possible to demonstrate experimental rigor.

One of the inherent drawbacks of the sequential ChIP is the loss of signal associated with the isolation of chromatin after the initial immunoprecipitation. This limitation can be circumvented somewhat with linear amplification of DNAs isolated in the end step. Recently, this has allowed for sequential ChIP to be used in combination with massive parallel sequencing to map epigenetic alterations associated with certain exposures and diseases [7]. Regardless, the efficiency of chromatin capture during immunoprecipitation will most likely dictate the chances of observing a reliable signal. As a result, liquid handling of the sample must be scaled up dramatically to achieve a robust and reliable signal. Larger-scale liquid handling presents its own disadvantages, particularly with regard to increased noise. The protocol outlined below is one that we have used and has produced reproducible and reliable results.

ChIP is a demanding technique and like most it requires care, perseverance, and attention to detail. It is not a functional assay. As a positive or negative indicator of transcription, ChIP is merely correlative. When specific areas of chromatin are enriched after precipitation with a specific antibody and quantified by PCR, ChIP is useful for increasing confidence regarding promoter or regulatory region occupancy by a given protein. The information gleaned from these types of experiments can provide important insights into alterations in chromatin configuration and the dynamic activities of transcriptional modifiers and regulators.

ChIP-re-ChIP can be a powerful technique, particularly when combined with massive parallel sequencing techniques. The protocol described below allows for reliable and reproducible results while offering flexibility for adjustment to different experimental conditions.

2 Materials

1. Harvest Buffer (10 ml): Tris/DTT, pH 8.3. Make fresh with DTT from a frozen stock. Add 1 ml of 1 M Tris, pH 8.3, 100 µl 1 M DTT, and 150 µl 10 N NaOH to 8.85 ml of deionized H_2O. Add the appropriate amount of protease inhibitor cocktail. Solution without protease inhibitor cocktail can be stored at 4 °C.
2. Lysis Buffer: 1% Triton X-100, 2 mM EDTA, 150 mM NaCl, and 20 mM Tris–HCl, pH 8.1, protease inhibitor cocktail. For 10 ml solution, add 1 ml of 10% Triton X-100, 40 µl of 0.5 M EDTA, 3.3 ml of 5 M NaCl, and 200 µl of 1 M Tris–HCl, pH 8.1 and 100 µl of protease inhibitor cocktail to 5.36 ml of deionized H_2O.
3. Dilution Buffer: 1% SDS, 10 mM EDTA, 50 mM Tris–HCl, pH 8.1, protease inhibitor cocktail. For a 10 ml solution, add 0.5 ml of 20% SDS, 200 µl EDTA, 1 ml 1% Triton X-100 and 0.5 ml of 1 M Tris–HCl, pH 8.1 and 100 µl of protease inhibitor cocktail to 3.2 ml of deionized H_2O.
4. Fixing Solution: 1% Formaldehyde. For 10 ml solution, add 3 ml of 37% formaldehyde to 7 ml 1 M HEPES buffer, pH 7.8. Make this solution fresh before the start of each experiment. Once chromatin has been fixed, the addition of protease cocktail inhibitor is no longer necessary; however, the addition of protease inhibitors will not interfere with the precipitation.
5. TSE I: 0.1% SDS, 1% Triton X-100, 2 mM EDTA, 20 mM Tris–HCl, pH 8.1, 150 mM NaCl. Solutions 5 and 6 are wash buffers of increasing stringency. They are made from the stock solutions listed above and can be made in large batches and stored at 4 °C.
6. TSE II: 0.1% SDS, 1% Triton X-100, 2 mM EDTA, 20 mM Tris–HCl, pH 8.1, 500 mM NaCl.
7. Buffer III: 0.25 M LiCl (from a 1 M solution), 1% IGEPAL CA-630, 1% deoxycholate, 1 mM EDTA, 10 mM Tris–HCl, pH 8.1. For 10 ml solution, add 250 µl of 1 M LiCl, 1 ml of 10% IGEPAL CA-630, 1 ml of 10% sodium deoxycholate, 20 µl 0.5 M EDTA and 100 µl of 1 M Tris–HCl, pH 8.1 to 7.63 ml of deionized H_2O.

8. TE8 (Tris–EDTA, pH 8.0): 10 mM Tris–HCl, 1 mM disodium EDTA, pH 8.0.
9. (An alternative lysis buffer) 0.5× RIPA Buffer: 0.1% SDS, 1% IGEPAL CA-630, 1% sodium deoxycholate, 25 mM 1 M Tris–HCl pH 7.6, and 150 mM NaCl, protease inhibitor cocktail. This can be stored at 4 °C.
10. Elution Buffer: 1% SDS, 0.1 M $NaHCO_3$.

3 Methods

3.1 Sequential ChIP Protocol: Overview

Generally, we use this protocol for a single ChIP, and scale it up 4–8 times for sequential ChIP depending on the entities being precipitated, their relative abundance and the cell line. This protocol was modified from the method of Ren and colleagues [8], which is a commonly used adaptation of the method described by Takahashi and colleagues [9].

Enough chromatin should be produced to use the two precipitation antibodies in both orders [10]. Alternatively, and particularly in the case where precipitating with an antibody directed to a protein that is not directly binding DNA, it can be useful to perform the ChIP with a DNA-binding moiety as the target of the primary precipitation [2] as typically is the case with transcription factors and co-regulatory proteins.

Four to eight 100-mm dishes of confluent cells are required for each precipitation. While a 250-mm dish might seem more expedient, we find that the volume of cells recovered from a 100-mm dish is more efficiently sonicated in 1.5 ml Eppendorf tube. So, while sonicating multiple tubes containing the cells from a 100-mm dish is more laborious and time-consuming, it is the only method we have found that enables one to perform a reliable sequential ChIP.

Proteins that directly bind to each other represent probably the *best-case* scenario for sequential ChIP. Proteins that do not directly interact with one another but are present together in a large multiprotein complex, may pose problems related to the integrity of the cross-linked complex. In addition, proteins that are present in a large variety of complexes may not be ideal targets either, as the target may represent only a small proportion of the total precipitate—though this is likely true for most proteins. Thus, chances of success targeting a DNA-binding protein as the primary precipitate will likely be higher than those targeting a non-DNA-binding moiety.

3.2 Preparation of Cross-Linked Chromatin

[All solutions should be ice cold.]

1. In a fume hood, fix cells by the addition of 1 ml of Fixing Solution (1% formaldehyde) per 10 ml of media in a 100-mm dish (eight 100-mm plates are sufficient for at least two

sequential ChIPs). Incubate for 10 min at room temperature. Formaldehyde is toxic to the optic nerve and should be handled with extreme care. The Fixing Solution should be used in the fume hood and media containing this solution should be disposed of in a suitable receptacle in the fume hood. *See* **Note 1** for alternative cross-linking strategies.

2. Aspirate media and wash cells twice with ice-cold PBS. At this stage, cells can be removed from the fume hood and work can be continued on the benchtop.
3. Aspirate PBS and harvest cells on each plate in 1 ml of Harvest Buffer. Adherent cells can be scraped off the dish using a Teflon cell scraper. Adherent cells should be centrifuged at 1200 × *g* in a swinging bucket rotor at 4 °C, washed twice in PBS and then the cell pellet can be resuspended in Harvest Buffer.
4. Transfer cell suspension to a 1.5 ml Eppendorf tube and incubate in a heating block at 30 °C for 15 min.
5. Spin cells for 5 min at 2000 × *g* (RCF). Wash with 1 ml ice-cold PBS and spin again for 5 min at 2000 × *g*. Aspirate PBS and resuspend each pellet in 600 μl of Lysis Buffer complete with protease inhibitors.
6. Sonicate each sample on ice three times for 10 s each. We typically have used a Branson 450 Model Sonifer but any similar model should be suitable. Standard settings are: Power Setting = 4, Duty Cycle = constant output, and Timer = Hold. Please *see* **Note 2** below regarding important considerations for the sonication of cells and the treatment of lysates.
7. Load 2 μl of lysate onto a 1% ethidium-stained agarose gel with a 1 kb marker and electrophorese the samples. The average size of the cross-linked chromatin DNA should be between 2 and 3 kb with no observable fluorescence above 3 kb or sample trapped in the wells (e.g., no unsheared fixed chromatin). If the chromatin DNA migrates as higher weight species, perform additional cycles of sonication until the desired length is observed. If chromatin DNA is significantly less that 2 kb, the sample should be abandoned and the assay re-started.
8. Once sonication is complete, centrifuge the samples at 12,000 × *g* at 4 °C for 15 min. Save 30 μl of each sample to be used as *Input* and determination of efficiency of co-immunoprecipitation.
9. To each *Input*, add 270 μl of Dilution Buffer complete with protease inhibitors and store at 4 °C.
10. Split the remaining sample into 2 aliquots (approx. 300 μl each). To each sample add 600 μl of Dilution Buffer complete with protease inhibitors. Each sample is suitable for the first precipitation with a single antibody (anti-X or anti-Y).

3.3 Clearing of Lysates

Prior to immunoprecipitation, samples must be cleared with agarose or Sepharose beads (*see* **Note 3**).

11. Wash the appropriate amount of Protein-A/G agarose with 10 volumes of TE8 buffer. Pulse centrifuge beads and aspirate TE8.
12. Repeat the same procedure as above 3–4 times.
13. Dilute the beads to a 25% slurry in Dilution Buffer and add 100 μl of this slurry to each sample along with 1–2 μg of goat-anti-rabbit IgG and 10 μg of sheared salmon sperm DNA (from a 1 mg/ml stock solution). Ideally, a final bead or resin volume of 15–25 μl is optimal. If DNA is used for high-throughput sequencing, poly dIdC should be used instead of sheared salmon sperm DNA. To make liquid handling easier with a large number of samples, the resin can be prepared as a large batch sufficient for the entire sample set.
14. Rock tubes in a cold room or chromatography refrigerator at 4 °C for 1 h.

3.4 Immunoprecipitation

15. Pulse spin lysates to pellet resin and transfer the aqueous lysate to a clean 1.5 ml Eppendorf tube.
16. Add 1–2 μg of affinity purified antibody directed to the protein of interest (*see* **Note 4**).
17. Precipitate chromatin on a rocker in a cold room or chromatography refrigerator at 4 °C overnight. It is important to note that the material is fixed. Thus, using antibodies that preferentially recognize native but not denatured protein may not work optimally for this assay. For this reason, many antibody manufacturers supply antibodies optimized for ChIP. We have used several such ChIP-verified antibodies from several different manufacturers with great success.
18. Add 100 μl of washed 25% slurry of the appropriate resin (protein-A agarose/sepharose in the case of rabbit IgG), 2 μg of salmon sperm DNA or poly dIdC and rock at 4 °C for 1 h (*see* **Note 5**).
19. Capture bead–antibody–chromatin complexes by centrifugation in a tabletop microfuge for approximately 10 s at 12,000 rpm (11,750 x *g*).
20. Aspirate supernatant and wash beads sequentially for 10 min each in TSE I, TSE II, and Buffer III (*see* **Note 6**). After each wash, pulse spin down the bead–chromatin complexes, aspirate supernatant and add the next buffer. Alternatively, we have also used 0.5× RIPA buffer in place of TSE I and TSE II without any noticeable decrease in signal. We find that washing is best performed using a soft Eppendorf tube holder adaptor

designed for the Vortex Genie™. We prefer to use setting 5 for all washes.

21. Wash the beads another three times for 10 s each in Tris–EDTA, pH 8.0.

3.5 Re-ChIP

22. For re-ChIP experiments, elute complexes from each sample of the primary ChIPs from the solid support resin (the beads) in 50 μl of 10 mM dithiothreitol (DTT) for 30 min at 37 °C. Sample replicates (4–6, we generally use 4) should then be pooled.
23. At this stage, the reducing agent DTT must be de-activated, removed, or diluted down below 5 mM or any antibody subsequently added to the mixture will be inactivated [11, 12]. We do not recommend de-salting using ion-exchange chromatography or other related methods as yields between samples tend to vary. We find that heat-inactivation (the 30 min, 37 °C incubation step), in combination with a simple 3–4× dilution in Dilution Buffer is sufficient for re-ChIP.
24. With the appropriate re-precipitation antibody, repeat **steps 15–21**.
25. Elute protein–DNA complexes in 300 μl of Elution Buffer at 65 °C overnight.
26. Extract protein–DNA complexes using 300 μl (or 1:1 volume) of phenol–chloroform–isoamyl alcohol 25:24:1. Add phenol–-chloroform–isoamyl alcohol solution. Vortex vigorously for 10–15 s and centrifuge samples at 12,000 × *g* for 10 min. Aspirate the aqueous fraction into a clean Eppendorf tube.
27. Precipitate DNA by the addition of 30 μl of 3 M sodium acetate solution, pH 5.2, 600 μl of ice-cold 100% ethanol and 2 μl of Pellet Paint®. The addition of Pellet Paint® (EMD Millipore) is optional and it may interfere with target amplification when performing quantitative Real-Time PCR.

Precipitated chromatin can be analyzed by PCR-agarose gel electrophoresis, quantitative real-time PCR, or it can be amplified and sequenced depending on the specific experimental end-point (*see* **Note** 7).

4 Notes

1. Although formaldehyde is highly recommended as a cross-linking reagent, a dual cross-linking protocol may be used. For dual cross-linking, prior to cross-linking with formaldehyde for 10 minutes, the cells are incubated with 1 mM dithiobis[succinimidylpropionate] (DSP) or 2 mM dimethyl 3.3-

dithiobisproprionamidate-2-HCl (DTBP) for 30 min at room temperature according to the manufacturer's instructions (Pierce) [13, 14].

2. Excessive sonication of the sample can lead to untoward heating of samples and disruption of chromatin complexes. The tip of the sonifier probe should never come in contact with the tube. By the same token, the probe must be the appropriate size and submerged as much as possible in the sample to prevent foaming. The sample should be ice cold and the tube should be firmly embedded in ice during sonication. Sonication can dramatically increase the temperature of the sample and extreme over heating of the sample can lead to degrading protein–protein and protein–DNA cross-links, reducing recovery.

   In order to reduce sonication duration and intensity, digestion of cross-linked chromatin with micrococcal nuclease (MNase) is often employed [15]. Generally, the chromatin is cleaved into mononucleosomes using 2.5 U of MNase/A_{260} (Worthington Biochemical Corporation) of nuclear suspension for 40 min at 37 °C [13]. We have used this protocol with varying degrees of success. MNase digestion requires an additional 1 h incubation and the MNase must be inactivated with EDTA. In addition, the amount of MNase and duration of digestion must be optimized for the amount of chromatin digested in each instance and this can be time-consuming and problematic when handling multiple samples.

3. The choice between agarose or Sepharose is largely dependent on the choice of the investigator. We have had success with both, but agarose is the most economical choice in most instances. As with any immunoprecipitation with a protein-conjugated resin, the choice of which protein-conjugated agarose/Sepharose is dependent on the species of the animal from which the precipitating antibody originated and the pre-clearing IgG should be from that species as well. For most applications and antibodies, either Protein-A or Protein-G conjugate is suitable.

4. Mouse, rabbit, and goat IgGs are not always suitable controls. We have found that some IgGs can be extremely noisy. A control IgG under stringent conditions should not precipitate any identifiable chromatin as assessed by polymerase chain reaction (PCR) amplification. If a control immunoglobulin is precipitating noticeable amounts of chromatin, other sources should be tested. For precipitation antibodies, one general rule of thumb is that an antibody that cannot efficiently immuno-deplete lysates of its target protein will not be suitable for ChIP. This may seem obvious but few experimentalists consider this before they spend long hours unsuccessfully precipitating the chromatin they desire. Another consideration, but not necessarily feasible in every situation, is that an antibody raised against a protein that

directly binds DNA will likely be more efficient than an antibody directed to a protein that binds DNA indirectly or through an ancillary protein.

5. In general, volumes of solid support matrices should be kept to a minimum. Agarose and Sepharose are notorious for adsorbing input but have high binding capacities when coupled to the appropriate IgG affinity matrix. Nonspecific binding naturally increases noise and requires blocking by one means or another. In the absence of poly-dIdC, resins can be pre-blocked and washed with bovine serum albumin. Protein-A binds to the Fc region of immunoglobulins through interactions with the heavy chain. The binding of protein A has been well documented for IgG from a variety of mammalian species and for some IgM and IgA as well. Protein-A Sepharose CL-4B (GE Healthcare Life Sciences) has a binding capacity of 20 mg human IgG/ml. Protein-G Sepharose 4 Fast Flow (GE Healthcare Life Sciences) should be selected for most mouse monoclonal antibodies because it binds to a broad range of IgG species and subclasses.
6. The stringency of washing conditions can be adjusted for any wash. Ultimately, this will determine the signal-to-noise ratio in the assay. We have always favored front-loading input and saving rigor and stringency for the culling of nonspecific material for the end of the process. In our experience, this has always seemed to favor the chances of success. For that reason, harsher washing conditions common to earlier protocols have been devoted to intermediate and later steps of the process.
7. Quantitative Real-Time PCR should be employed to quantify the amount of chromatin corresponding to regions of interest solely to increase confidence concerning significant differences in enrichment from one condition to another; little else can be inferred from this information. DNA recovered from eluted and denatured chromatin should be quantified (A_{260}) and this can be used to inform parameters for PCR, but it is likely that starting amounts for amplification will need to be determined by trial and error for every experiment.

References

1. Beischlag TV, Wang S, Rose DW, Torchia J, Reisz-Porszasz S, Muhammad K et al (2002) Recruitment of the NCoA/SRC-1/p160 family of transcriptional coactivators by the aryl hydrocarbon receptor/aryl hydrocarbon receptor nuclear translocator complex. Mol Cell Biol 22:4319–4333
2. Labrecque MP, Takhar MK, Jagdeo JM, Tam KJ, Chiu C, Wang TY et al (2014) A TRIP230-retinoblastoma protein complex regulates hypoxia-inducible factor-1alpha-mediated transcription and cancer cell invasion. PLoS One 9(6):e99214
3. Parekh BS, Maniatis T (1999) Virus infection leads to localized hyperacetylation of histones H3 and H4 at the IFN-beta promoter. Mol Cell 3:125–129
4. Wathelet MG, Lin CH, Parekh BS, Ronco LV, Howley PM, Maniatis T (1998) Virus infection induces the assembly of coordinately activated

transcription factors on the IFN-beta enhancer in vivo. Mol Cell 1:507–518
5. Shang Y, Hu X, DiRenzo J, Lazar MA, Brown M (2000) Cofactor dynamics and sufficiency in estrogen receptor-regulated transcription. Cell 103:843–852
6. Shang Y, Myers M, Brown M (2002) Formation of the androgen receptor transcription complex. Mol Cell 9:601–610
7. Shankaranarayanan P, Mendoza-Parra MA, van Gool W, Trindade LM, Gronemeyer H (2012) Single-tube linear DNA amplification for genome-wide studies using a few thousand cells. Nat Protoc 7:328–338
8. Ren B, Cam H, Takahashi Y, Volkert T, Terragni J, Young RA et al (2002) E2F integrates cell cycle progression with DNA repair, replication, and G(2)/M checkpoints. Genes Dev 16:245–256
9. Takahashi Y, Rayman JB, Dynlacht BD (2000) Analysis of promoter binding by the E2F and pRB families in vivo: distinct E2F proteins mediate activation and repression. Genes Dev 14:804–816
10. Beischlag TV, Perdew GH (2005) ER alpha-AHR-ARNT protein-protein interactions mediate estradiol-dependent transrepression of dioxin-inducible gene transcription. J Biol Chem 280:21607–21611
11. Cleland WW (1964) Dithiothreitol, a new protective reagent for Sh groups. Biochemistry 3:480–482
12. Olson PR, Weiblen BJ, O'Leary JJ, Moscowitz AJ, McCullough J (1976) A simple technique for the inactivation of IgM antibodies using dithiothreitol. Vox Sang 30:149–159
13. Drobic B, Perez-Cadahia B, Yu J, Kung SK, Davie JR (2010) Promoter chromatin remodeling of immediate-early genes is mediated through H3 phosphorylation at either serine 28 or 10 by the MSK1 multi-protein complex. Nucleic Acids Res 38:3196–3208
14. Khan DH, Gonzalez C, Cooper C, Sun JM, Chen HY, Healy S et al (2014) RNA-dependent dynamic histone acetylation regulates MCL1 alternative splicing. Nucleic Acids Res 42:1656–1670
15. Carey MF, Peterson CL, Smale ST (2009) Chromatin immunoprecipitation (ChIP). Cold Spring Harb Protoc 2009:pdb prot5279

Chapter 10

Sm-ChIPi: Single-Molecule Chromatin Immunoprecipitation Imaging

Roubina Tatavosian and Xiaojun Ren

Abstract

Epigenetic complexes regulate chromatin dynamics *via* binding to and assembling on chromatin. However, the mechanisms of chromatin binding and assembly of epigenetic complexes within cells remain incompletely understood, partly due to technical challenges. Here, we present a new approach termed single-molecule chromatin immunoprecipitation imaging (Sm-ChIPi) that enables to assess the cellular assembly stoichiometry of epigenetic complexes on chromatin. Sm-ChIPi was developed based on chromatin immunoprecipitation followed by single-molecule fluorescence microscopy imaging. In this method, an epigenetic complex subunit fused with a gene coding for a fluorescent protein is stably expressed in its corresponding knockout cells. Nucleosomes associated with epigenetic complexes are isolated from cells at native conditions and incubated with biotinylated antibodies. The resulting complexes are immobilized on a quartz slide that had been passivated and functionalized with NeutrAvidin. Image stacks are then acquired by using single-molecule TIRF microscopy. The individual spots imaged by TIRF microscopy represent single protein–nucleosome complexes. The number of copies of the protein complexes on a nucleosome is inferred from the fluorescence photobleaching measurements. Sm-ChIPi is a sensitive and direct method that can quantify the cellular assembly stoichiometry of epigenetic complexes on chromatin.

Key words Epigenetics, Epigenetic complex, Chromatin, Polycomb group (PcG) proteins, PRC2, PRC1, Nucleosome, Chromatin-binding mechanism, Epigenetic protein–nucleosome complex, Single-molecule fluorescent microscopy

1 Introduction

Nucleosome, the basic building unit of chromatin, is a histone octamer consisting of two copies of each of H2A, H2B, H3, and H4, around which approximately 147 bp of DNA is wrapped [1]. The histone tails and, to a lesser extent, the globular domains of histones are subject to post-translational modifications by epigenetic protein complexes (*writer complexes*) [2, 3]. These modifications either directly regulate chromatin structure or recruit epigenetic protein complexes (*reader complexes*) that impact genome organization and regulate gene transcription [2, 3]. This is exemplified by Polycomb group (PcG) proteins [4–12].

Neus Visa and Antonio Jordán-Pla (eds.), *Chromatin Immunoprecipitation: Methods and Protocols*,
Methods in Molecular Biology, vol. 1689, DOI 10.1007/978-1-4939-7380-4_10, © Springer Science+Business Media LLC 2018

PcG proteins mediate epigenetic inheritance of repressive states of transcription [4–18]. However, the molecular mechanisms by which epigenetic protein complexes assemble on and spread along chromatin are poorly understood.

Currently, the standard methods for determining interactions between epigenetic protein complexes and chromatin is chromatin immunoprecipitation (ChIP) followed by quantitative PCR (ChIP-qPCR) or high-throughput sequencing (ChIP-seq). Sequential ChIP followed by mass spectrometry has also been developed to assess the combinatorial histone modifications within a purified nucleosome [19, 20]. However, the quantification of the cellular stoichiometry of epigenetic protein complexes on native nucleosomes poses a formidable challenge.

Single-molecule approaches based on molecular counting of fluorescent proteins/fluorophores have been widely applied to investigate the molecular interactions of macromolecular complexes [21–26]. ChIP followed by single-molecule counting and imaging has been developed to quantify the combinatorial epigenetic states within a native mononucleosome [27, 28]. Here, we describe the details of a novel approach termed single-molecule ChIP imaging (Sm-ChIPi) that is based on ChIP followed by single-molecule fluorescence imaging [17]. Sm-ChIPi can quantify the molecular stoichiometry of epigenetic protein complexes on chromatin at their native states. We have verified Sm-ChIPi using epigenetic protein–nucleosome complexes with a known stoichiometry and applied Sm-ChIPi to investigate the cellular stoichiometry of PRC1 and PRC2 on chromatin [17].

2 Materials

Prepare all the solutions with autoclaved Milli-Q water.

2.1 Preparation of Native Nucleosomes from Mammalian Cells

1. 10× citrate saline solution: 135 mM potassium chloride, 15 mM sodium citrate.
2. Washing buffer: sterile phosphate saline buffer (PBS), pH 7.4.
3. Crosslinking buffer: 4% paraformaldehyde (PFA) stock solution. Heat 500 ml PBS pH 7.4 to 62–64 °C. Add 80 μl of 5 N NaOH and 20 g PFA. While maintaining temperature at the range of 62–64 °C, stir the mixture until it gradually clears out. The pH of the fixative should be 7.4. Filter the solution and store it at −20 °C.
4. Buffer A:10 mM HEPES, pH 7.9, 1.5 mM $MgCl_2$, 1.0 mM Na_3VO_4, 10 mM KCl, 340 mM sucrose, 10% glycerol, 50 μg/ml bovine serum albumin, 1× protease inhibitor cocktail, 0.1 mM PMSF.
5. Buffer B: Buffer A plus 0.2% Triton X-100.

6. Buffer C: Buffer A plus 30% (W/V) sucrose.
7. UV–Vis spectrophotometer.
8. 2 M NaCl solution.
9. Micrococcal nuclease (MNase).
10. 1 mM $CaCl_2$.
11. 4 mM EGTA, pH 8.0.
12. TE buffer: 10 mM Tris–HCl, pH 8.0, 1 mM EDTA.
13. 20 mg/ml RNase A.
14. 20 mg/ml Proteinase K.
15. Phenol:chloroform:isoamyl alcohol (25:24:1).
16. Chloroform.
17. 3 M sodium acetate, pH 5.2.
18. Ethanol 200 Proof.

2.2 Linear Sucrose Gradient Ultracentrifugation

1. TH-641 swinging bucket rotor and Sorvall WX ultracentrifuge.
2. Thick wall polyallomer tubes (PA, 9/16 × 3 1/2) (Beckman Coulter, 331372).
3. Sucrose Buffer M: 10 mM HEPES, pH 7.9, 50 μg/ml BSA, 10 mM KCl, 1.5 mM EDTA, 1.0 mM Na_3VO_4, 0.2 mM DTT, 0.5 mM PMSF.

2.3 Construction and Passivation of Flow Chambers

1. Quartz slides.
2. Coverslip.
3. Four vertical staining jars with covers.
4. Bath sonicator.
5. 1 M potassium hydroxide.
6. Aminosilane.
7. Propane torch.
8. Epoxy glue.
9. Glacial acetic acid.
10. mPEG-SVA (Laysan Bio, Inc., MPEG-SVA-5000).
11. Biotin PEG-SVA (Laysan Bio, Inc., 256–586-9004).
12. 10 mM sodium bicarbonate, pH 8.5.
13. Methanol.
14. Double-sided tape.
15. Nitrogen gas for storage of the reagents and flow chambers.
16. T50 buffer: 10 mM Tris–HCl, pH 8.0, 50 mM NaCl. Store at 4 °C up to 2 months.
17. T50-BSA buffer: T50 buffer with 0.1 mg/ml BSA. Store at 4 °C up to 2 months.

2.4 Immuno-precipitation of Epigenetic Protein-Nucleosome Complexes

1. Biotinylated antibody (anti-H2A, H2B, H3, H4; or anti-GFP, etc.).
2. NeutrAvidin: Prepare stock solution of 0.2 μg/ml using pre-chilled T50 buffer.

2.5 Single-Molecule Total Internal Reflection Fluorescent Imaging

1. Zeiss Axio Observer D1 Manual Microscope (Zeiss, Germany).
2. Alpha Plan-Apochromatic 100×/1.46 NA Oil immersion objective (Zeiss, Germany).
3. Evolve 512 × 512 EMCCD camera (Photometric, Tucson, AZ).
4. Excitation laser 488; 1.2 mV (Intelligent Imaging Innovations, CO).

3 Method

3.1 Preparation of Native Nucleosomes from Mammalian Cells

To minimize photobleaching of fluorescent protein, all glass containers that hold samples are wrapped with aluminum foil at all the time.

1. Harvest approximately 5 × 10^8 adherent mammalian cells from 15-cm dish using pre-warm (1×) citrate saline solution (*see* **Note 1**). For non-adherent cells, *see* **Note 2**.
2. Collect the cell pellets by centrifugation at 300 × *g* for 5 min at 4 °C.
3. Wash the cell pellets with cold PBS twice and each time collect the pellets by centrifugation at 300 × *g* for 5 min at 4 °C.
4. To preserve epigenetic protein–nucleosome complexes, add 4 ml of pre-chilled 2% PFA directly to the cell pellets and incubate on ice for 10 min.
5. Centrifuge the sample at 300 × *g* for 5 min at 4 °C to collect cell pellets. Decant the supernatant and assess the packed cell pellet volume (PCV).
6. Gently resuspend the pellet in 2.5 PCV of cold Buffer A until the pellet is completely homogenized.
7. Add 2.5 PCV of cold Buffer B and gently mix the solution by inverting 4–6 times. Incubate on ice for 10 min with occasional gentle inversions every 2–3 min.
8. Centrifuge at 1300 × *g* for 5 min at 4 °C. Isolate the pellets which contain cell nuclei and discard the supernatant.
9. Gently resuspend the nuclei pellets with 6 × PCV of buffer A to obtain a homogeneous solution.

10. Prepare 7.5 ml of pre-chilled buffer C. Slowly and gently apply the resuspended nuclei pellets to the top of the buffer C (*see* **Note 3**).
11. Collect the pellets (pure nuclei) by centrifugation at 1300 × *g* for 12 min at 4 °C.
12. Gently resuspend the pellets in 2× PCV of pre-chilled buffer A while holding the sample on ice. Transfer the sample into 1.5 ml centrifuge tube.
13. To quantify concentration of DNA within the isolated nuclei (*see* **Note 4**), add 10 μl of the resuspended nuclei (from **step 12**) to 500 μl of 2 M NaCl and sonicate it for 5 min at 40% amplitude, 30 s ON, and 30 s OFF settings.
14. Use UV–Vis spectroscopy to measure and to back calculate concentration of DNA at 260 nm wavelength (*see* **Note 5**).
15. The final concentration of nuclei pellet should be 2.0 μg/ml. If necessary, dilute the isolated nuclei pellet (from **step 11**) using pre-chilled buffer A.
16. Add $CaCl_2$ to the sample to achieve a final concentration of 1 mM.
17. To generate mononucleosomes, digest the nuclei pellet with 1.4 U/ml MNase for 8 min at 37 °C. To produce polynucleosomes, digest it with 0.7 U/ml for 8 min at 37 °C (*see* **Note 6**).
18. To stop the digestion reaction, immediately add pre-chilled EGTA to final working concentration of 4 mM.
19. Recover the supernatant by centrifugation at 20,000 × *g* for 5 min at 4 °C (*see* **Note 7**). The supernatant contains the fragmented nucleosomes.
20. Use 50 μl of digested chromatin sample to confirm the size of fragmented chromatin by extraction of nucleosomal DNA. Add 50 μl of TE buffer to the sample.
21. Nucleosomal DNA of the fragmented chromatin should be extracted to confirm the size of recovered nucleosomes. Aliquot 50 μl of digested chromatin sample and mix it with 50 μl of TE buffer.
22. Reverse crosslink the epigenetic protein–nucleosome complexes by incubating at 65 °C for 1 h.
23. To digest proteins and RNAs, add 2.0 μl of 20 mg/ml proteinase K and 2.5 μl of 20 mg/ml RNase A to the solution. Mix and incubate at 55 °C for 3–4 h.
24. Add equal volume of phenol:chloroform to the sample, and vortex vigorously for 2 min to isolated DNA (*see* **Note 8**). Carefully transfer the top layer (aqueous layer) to a new tube.

25. Add a volume of chloroform equal to volume of top layer extracted from the above step and vortex for 5 min. Transfer the top layer (aqueous layer) to a new tube.
26. To precipitate DNA out of the solution, add 10% volume of 3 M sodium acetate pH 5, 2.5 μl glycogen 20 mg/ml (*see* **Note 9**), and 2.5 volume ethanol. After addition of each reagent, vortex it thoroughly. DNA might not precipitate out completely from the solution without vortexing.
27. Place the sample at −20 °C overnight to precipitate DNA out of the solution.
28. To collect the DNA pellets, centrifuge the tubes at 20,000 × *g* for 30 min at 4 °C. Carefully decant the supernatant and dry the DNA pellets at room temperature for 5–10 min.
29. Dissolve DNA pellets in 20 μl of autoclaved Milli-Q water.
30. Run 1.2–2% agarose gel to confirm the sizes of nucleosome fragments (*see* **Note 10**).

3.2 Linear Sucrose Gradient Ultracentrifugation

1. To further purify mononucleosomes, use pre-chilled buffer M and applicable amount of sucrose to prepare 5–30% sucrose gradient solutions with increments of 5% (e.g., 5%, 10%, etc.). To isolate larger nucleosomal fragments (polynucleosomes) prepare 15–40% sucrose gradient solutions. Set up sucrose gradient by gradually layering sucrose solutions in the pre-chilled ultracentrifuge tube by increasing sucrose solution density from bottom to the top of the test tube (*see* **Note 11**) (Fig. 1).
2. Dilute the sample with highest sucrose gradient concentration so that the final concentration of the sucrose in the sample reaches approximately equivalent to the sucrose concentration of the very top layer. The sample should contain 200–300 μg of DNA and the total volume should be 0.5 ml (*see* **Note 12**).
3. Carefully add the diluted nucleosomal solution to the top of the last sucrose gradient layer. The tube should be handled and loaded into the rotor carefully to minimize disruption of the sucrose gradient layers.
4. Centrifuge the sample at 200,000 × *g* for 18–20 h at 4 °C (*see* **Note 13**).
5. Fractionate the sample(s) by gently collecting 0.5 ml of sample at a time, starting from the top of the centrifuge tube using a 1000 μl pipette.
6. Analyze nucleosomal DNA fragments from each fraction using DNA extraction protocol as previously explained in Subheading 3.1, **steps 21–30**.
7. Identify the fractions that you intend to use for Sm-ChIPi and move to Subheading 3.4.

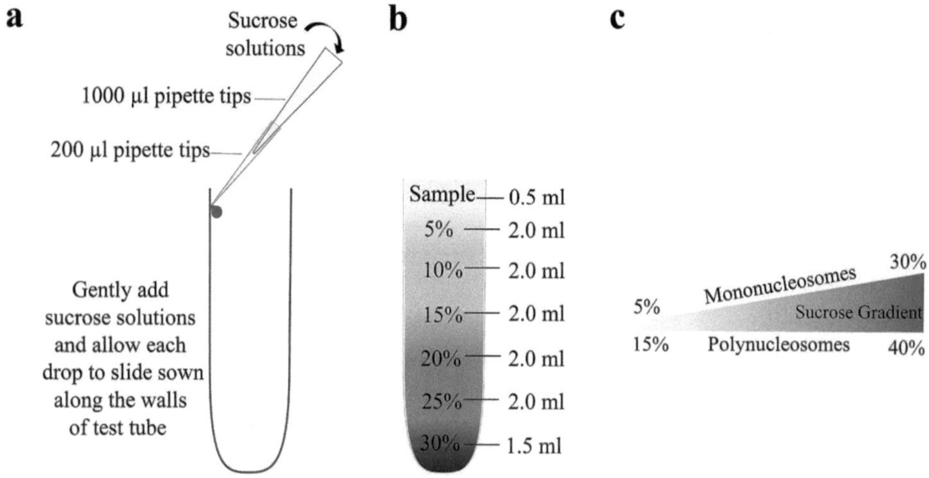

Fig. 1 Preparation of sucrose gradient. (**a**) Insert the P-200 pipettor tip at the end of P-1000 pipettor tip to transfer sucrose solutions into polyallomer tube while holding it straight up. Every drop has to be added very gently and in a dropwise manner by applying each drop on the walls of the tube and letting gravity pull the solution down. (**b**) The figure illustrates the volume to use for each of the sucrose density fractions to create a 5–30% gradient. (**c**) To obtain mononucleosome, prepare 5–30% sucrose gradient. To generate polynucleosomes fragments, a 15–40% sucrose gradient is required

3.3 Construction and Passivation of Flow Chambers

Carry out all the steps at room temperature unless otherwise specified. Use Milli-Q water for all the rinsing and washing steps. This protocol was adopted from [29].

1. Drill two even sized holes in a quartz slide, about 0.8 mm in diameter, across from each other (Fig. 2a).
2. Wash the slide with water and detergent 2–3 times. At the last wash, rinse the slide with water to remove any detergent residues.
3. Arrange two sets of coverslips (A and B) for each flow chamber preparation. Coverslip A will be used for the passivation reaction and coverslip B will be used for sandwiching the passivated coverslip A (*see* **Note 14**).
4. Sonicate the slide and both the coverslip sets (A and B) with water in the separate slide holders for 15 min.
5. Rinse the slide and both the coverslip sets with water 2–3 times. Further sonicate them with methanol for additional 15 min (*see* **Note 15**). Quickly rinse the slide and coverslip B with methanol and incubate them in methanol until the flow chamber assembly time (B coverslip will be used in the **step 10** below).
6. Sonicate coverslip A in 1 M KOH for 40 min to activate the glass surface for aminosilane functionalization. Rinse the coverslip A a few times with water.

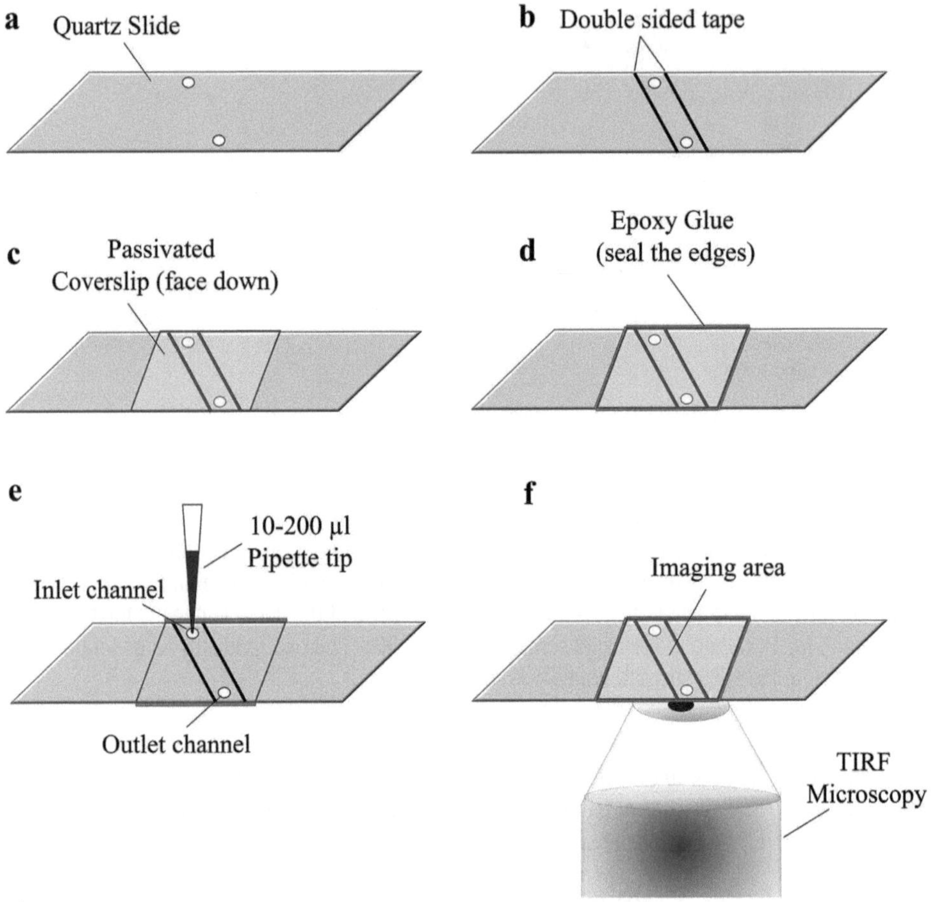

Fig. 2 Construction of flow chamber. (**a**) To assemble the flow chamber, drill two holes across from each other in a quartz slide. (**b**) Use two pieces of double-sided tape to create a channel around the holes. (**c**) Assembles the passivated coverslip by placing the passivated side down on the slide. (**d**) Use epoxy glue to completely seal the edges of the coverslip. (**e**) Use 10 or 200 μl pipettes to flow solutions through the chamber using the inlet/outlet holes located on the slide. (**f**) Use TIRF microscopy to acquire image stacks

7. Dry the coverslip A with nitrogen gas. To completely remove the remaining moisture, burn the coverslip with propane torch for about 1–2 s each side and immediately place it in a pre-dried slide holder (*see* **Note 16**).
8. Add 95 ml ethanol, 5 ml acetic acid, and 1 ml aminosilane respectively to the pre-dried slide holder containing coverslip A and incubate it for 20 min in dark.
9. Rinse coverslip A first with methanol and then with water, each time for about 1–2 min. Dry coverslip A with nitrogen gas and store it in a humidified box.
10. Dry coverslip B and slide with nitrogen gas and store it in a dry slide holder.

11. For each coverslip, mix 16 mg of mPEG-SVA and 0.3 mg of biotin-PEG. Dissolve the dry reagents in 70 μl of freshly prepared 10 mM Sodium bicarbonate pH (8.5) per coverslip (*see* **Note 17**).
12. Gently apply 70 μl of the above solution to coverslip A and immediately sandwich it with coverslip B. Incubate for 4 h in the humidified box in dark.
13. Separate coverslip A from coverslip B and rinse coverslip A with water for 1–2 min. Discard coverslip B. Dry coverslip A with nitrogen gas and place it on a dust free tissue paper facing passivated side up.
14. Use nitrogen gas to dry the clean slide. Prepare a flow chamber by sandwiching the slide and the passivated coverslip A with two pieces of double-sided tape (*see* **Note 18**), forming a channel about 0.3 mm from each sides of the inlet/outlet holes (Fig. 2b, c).
15. The edges of the sandwich should be carefully sealed with epoxy glue and allowed to dry for 5–10 min at room temperature (Fig. 2d). The flow chamber can hold approximately 10–15 μl of aqueous solution. The flow chamber can be stored at −20 °C in nitrogen gas in the dark up to 3 weeks.
16. The quality of flow chamber should be tested prior to use by adding of 100 μl of T50 buffer into the chamber via inlet hole and letting the buffer flow out from the outlet hole (Fig. 2e). Use TIRF microscopy to visualize fluorescent spots (Fig. 2f). Determine the amount of contribution of the background fluorophores by directly counting the visible spots within the imaging surface (*see* **Note 19**).
17. To determine the contribution of nonspecific binding proteins on the passivated surface to the overall fluorophore spots, add 50 μl of fractionated lysate (epigenetic protein–nucleosome complexes) without antibody. Incubate it for 10 min in dark at 4 °C. Remove unbound lysate by flowing 100 μl T-50 buffer. Use TIRF microscopy to quantify fluorophore spots (*see* **Note 20**).

3.4 Immunoprecipitation of Epigenetic Protein–Nucleosome Complexes

All the reagents and buffers for this section are pre-chilled prior to use. All the incubations are performed at 4 °C. All procedures are performed in dark to minimize photobleaching during the experimental set up.

1. Dilute biotinylated antibody with pre-chilled T50-BSA buffer. Concentration of anti-histone antibody should be within the range of 100–200 nM. For antibodies that target genetically encoded fluorescent protein tags within the epigenetic complex, use final concentration of 5–10 nM.

2. Prepare the immobilization complex (epigenetic protein–nucleosome–antibody) by mixing 35 μl of protein fraction that intend to be analyzed (*see* **Note 21**), 1 μl of diluted biotinylated antibody (*see* **Note 22**), and 64 μl of T50 buffer. Incubate at 4 °C for 2–12 h.
3. Crosslink antibody-epigenetic protein–nucleosome complex with 2% PFA to the final working concentration of 0.5% for 15 min at 4 °C.
4. Prepare 0.2 μg/ml stock solution of NeutrAvidin with T50 buffer and keep the solution in the dark at 4 °C.
5. Follow the steps below to prepare the passivated flow chamber for TIRF imaging.
   (a) Flow 100 μl of T50-BSA buffer to minimize binding of nonspecific proteins to the surface.
   (b) Add 50 μl of 0.2 μg/ml NeutrAvidin and incubate on ice and in the dark for 15 min.
   (c) Wash the flow chamber with 100 μl of T50 buffer to remove excess of NeutrAvidin residues.
   (d) Add the crosslinked immobilization complex to the flow chamber and incubate for 10 min.
   (e) Gently flow 100 μl of T50 buffer to remove the floating and unbound complex from the flow chamber.
   (f) Image the sample using TIRF microscopy (Fig. 3).

3.5 Single-Molecule Total Internal Reflection Fluorescent Imaging

1. Use total internal fluorescent (TIRF) microscopy to acquire image stacks. Under our experimental conditions, 200 frames were acquired at 100 frames per millisecond and 200 ms integration time (*see* **Note 23**). To eliminate the possibility of missing a bleaching step, it is important to minimize light exposure to the sample during the imaging preparation period.
2. To improve dye stability during Sm-ChIPi, oxygen scavenger reagents can be added to the immobilized complex immediately before imaging.

3.6 TIRF Data Analysis

1. Use ImageJ (http://imagej.nih.gov/ij/) to generate fluorescent intensity time trace of the discrete points from the acquired image stacks (*see* **Note 24**).
2. Detect and identify photobleaching steps by Chang-Kennedy filter [30]. Each photobleaching step corresponds to the number of fluorophores associated with protein present in the immobilization complex. For example, if protein that is subject to Sm-ChIPi analysis exist in a monomeric form, one photobleaching step in the time trace of fluorescent intensity would

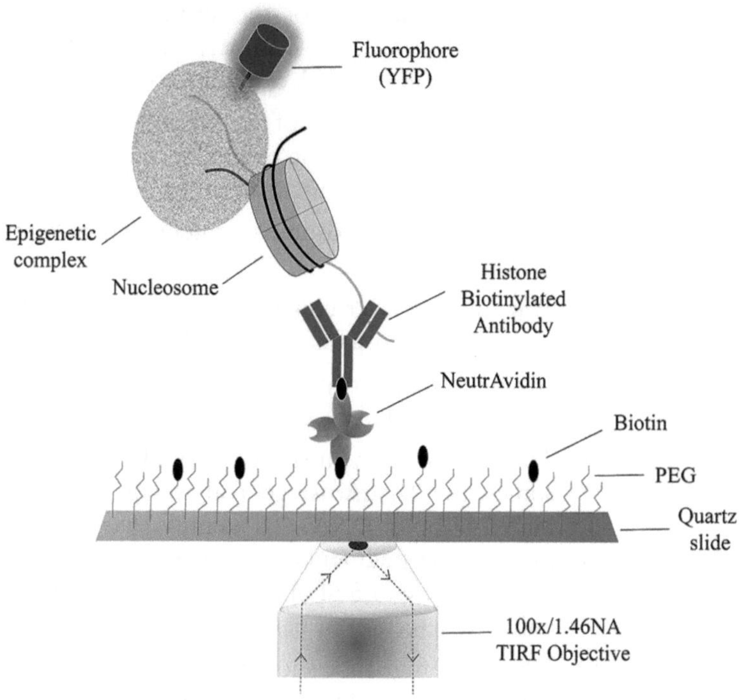

Fig. 3 The figure provides a schematic representation of the immobilization and imaging of an epigenetic protein–nucleosome complex. The sample is immobilized on a passivated, NeutrAvidin-coated flow chamber via a biotinylated anti-histone antibody. TIRF microscopy is used to detect and quantify the fluorophores

be identified. Likewise, if there are four fluorophores associated with the protein complex, they will bleach in a sequential order, thus, four photobleaching steps will be detected.

4 Notes

1. Use pre-warmed sterile PBS to dilute citrate saline.
2. For non-adherent mammalian cells, skip the first step and move on to **step 2**.
3. The purpose of this step is to isolate pure nuclei by passing it through a high concentration of sucrose gradient. Use 15 ml test tube for this part of the procedure. The resuspended nuclei solution should be added very gently to the top layer of the sucrose cushion. Allow some mixing to occur at the interface otherwise the surface tension of the sucrose cushion may prevent nuclei from passing through it.
4. To achieve desired nucleosome fragment size, it is critical to determine the DNA concentration from the isolated nuclei pellet prior to addition of MNase enzyme.

5. DNA concentration of the suspended nuclei can be back-calculated using Beer–Lambert's law. Multiply the measured concentration by DNA extinction coefficient value ($\varepsilon = 50$).
6. The enzymatic activity of the MNase is defined based on Sigma units. After the addition of MNase, mix the solution gently and thoroughly prior to incubation.
7. The digested nucleosome fragments can be kept at −20 °C for long-term storage.
8. The aqueous layer contains DNA fragments. Proteins will be trapped in the organic layer of phenol:chloroform.
9. Glycogen helps to precipitate DNA out of solution and will increase the DNA recovery yield.
10. To confirm fragment size of isolated DNA from mononucleosomes, use 2% agarose gel (on average each nucleosome contains 147 bp of DNA). To identify DNA obtained from larger nucleosome fragments use 1.2–1.5% agarose gel. We recommend the use of 5 μl of 80% glycerol instead of loading dye to enhance detection of bands under the UV light.
11. Ultracentrifuge tube should be held up right while it is securely placed in the ice. Insert the P-200 pipettor tip at the end of P-1000 pipettor tip to transfer sucrose solutions into polyallomer tube very gently and in a dropwise manner. Apply each drop on the walls of the tube and let gravity pull the solution down (avoid dropping the sucrose solution directly). Discrete layers of different sucrose gradients should be visible.
12. Using higher concentration of DNA will result in poor separation of nucleosomes.
13. Longer acceleration rate and deceleration rate is recommended if swinging bucket rotor is used for the density gradient centrifugation to minimize the disturbance of the sucrose gradient layers.
14. Each slide holder has capacity to fit eight sliders/coverslips, hence, up to eight flow chamber can be prepared at once.
15. Methanol removes organic residues from the surface of slides and coverslips.
16. Since aminosilane is highly photosensitive and hydrolyzes quickly in water, coverslips and the slide holder have to be completely moisture free prior to use for the aminosilane reaction. To completely remove moistures from the slide holder, place the slide holder and its cover in drying oven overnight, while they are completely wrapped with aluminum foil (since this reaction needs to be performed in dark, the slide holder must be wrapped in a dry aluminum foil). Take them out of the oven 3–4 h prior to use and store them immediately in a

desiccator to maintain dryness. The aminosilane reaction has to be carried out at room temperature. Aminosilane has to be stored in dark under nitrogen gas.

17. mPEG has half-life of about 10 min in pH 8.5. Therefore, it is critical to use it immediately after addition of sodium bicarbonate.
18. Exert small amount of force to the coverslip to ensure that the double-sided tape completely sticks to both coverslip and slide.
19. Under our experimental conditions, the contribution of the visible fluorophores on the blank passivated flow chamber was about 0.5% (~5 spots) per 1000 μm^2 imaging area.
20. Under our experimental conditions, the involvement of the nonspecifically absorbed fractionated lysate on the flow chamber was less than 1%.
21. If necessary, lysate and/or antibody can be further diluted with T50-BSA buffer to maintain 300–600 points per 1000 μm^2 imaging area.
22. Biotinylated antibody can be either against one of the histone subunits (H2A, H2B, H3, and H4) or against fluorescent protein that is genetically fused into the protein of interest.
23. Frame number and the integration time may be adjusted based on the life time of the fluorescent protein that is used to tag the epigenetic protein. In order to assess accurate counting of the photobleaching steps, adjust the frame number so that at least 90% of the fluorophores are bleached at the last frame.
24. Check all acquired movies and discard any movies or data points that shifted or photo-blinked during imaging.

Acknowledgment

This work was supported by the National Cancer Institute of the National Institutes of Health under Award Number R03CA191443 (to X.R.).

References

1. Luger K, Dechassa ML, Tremethick DJ (2012) New insights into nucleosome and chromatin structure: an ordered state or a disordered affair? Nat Rev Mol Cell Biol 13:436–447
2. Jenuwein T, Allis CD (2001) Translating the histone code. Science 293:1074–1080
3. Li B, Carey M, Workman JL (2007) The role of chromatin during transcription. Cell 128:707–719
4. Kerppola TK (2009) Polycomb group complexes—many combinations, many functions. Trends Cell Biol 19:692–704
5. Simon JA, Kingston RE (2013) Occupying chromatin: polycomb mechanisms for getting to genomic targets, stopping transcriptional traffic, and staying put. Mol Cell 49:808–824
6. Di Croce L, Helin K (2013) Transcriptional regulation by polycomb group proteins. Nat Struct Mol Biol 20:1147–1155

7. Aranda S, Mas G, Di Croce L (2015) Regulation of gene transcription by polycomb proteins. Sci Adv 1:e1500737
8. Blackledge NP, Rose NR, Klose RJ (2015) Targeting polycomb systems to regulate gene expression: modifications to a complex story. Nat Rev Mol Cell Biol 16:643–649
9. Kondo T, Ito S, Koseki H (2016) Polycomb in transcriptional phase transition of developmental genes. Trends Biochem Sci 41:9–19
10. Margueron R, Reinberg D (2011) The polycomb complex PRC2 and its mark in life. Nature 469:343–349
11. Steffen PA, Ringrose L (2014) What are memories made of? How polycomb and trithorax proteins mediate epigenetic memory. Nat Rev Mol Cell Biol 15:340–356
12. Schwartz YB, Pirrotta V (2013) A new world of polycombs: unexpected partnerships and emerging functions. Nat Rev Genet 14:853–864
13. Ren X, Vincenz C, Kerppola TK (2008) Changes in the distributions and dynamics of polycomb repressive complexes during embryonic stem cell differentiation. Mol Cell Biol 28:2884–2895
14. Ren X, Kerppola TK (2011) REST interacts with Cbx proteins and regulates polycomb repressive complex 1 occupancy at RE1 elements. Mol Cell Biol 31:2100–2110
15. Cheng B, Ren X, Kerppola TK (2014) KAP1 represses differentiation-inducible genes in embryonic stem cells through cooperative binding with PRC1 and derepresses pluripotency-associated genes. Mol Cell Biol 34:2075–2091
16. Zhen CY, Duc HN, Kokotovic M, Phiel CJ, Ren X (2014) Cbx2 stably associates with mitotic chromosomes via a PRC2- or PRC1-independent mechanism and is needed for recruiting PRC1 complex to mitotic chromosomes. Mol Biol Cell 25:3726–3739
17. Tatavosian R, Zhen CY, Duc HN, Balas MM, Johnson AM, Ren X (2015) Distinct cellular assembly stoichiometry of polycomb complexes on chromatin revealed by single-molecule chromatin immunoprecipitation imaging. J Biol Chem 290:28038–28054
18. Zhen CY, Tatavosian R, Huynh TN, Duc HN, Das R, Kokotovic M, Grimm JB, Lavis LD, Lee J, Mejia FJ, Li Y, Yao T, Ren X (2016) Live-cell single-molecule tracking reveals co-recognition of H3K27me3 and DNA targets polycomb Cbx7-PRC1 to chromatin. Elife 5. doi:10.7554/eLife.17667
19. Voigt P, LeRoy G, Drury WJ 3rd, Zee BM, Son J, Beck DB, Young NL, Garcia BA, Reinberg D (2012) Asymmetrically modified nucleosomes. Cell 151:181–193
20. Sadeh R, Launer-Wachs R, Wandel H, Rahat A, Friedman N (2016) Elucidating combinatorial chromatin states at single-nucleosome resolution. Mol Cell 63:1080–1088
21. Jain A, Liu R, Ramani B, Arauz E, Ishitsuka Y, Ragunathan K, Park J, Chen J, Xiang YK, Ha T (2011) Probing cellular protein complexes using single-molecule pull-down. Nature 473:484–488
22. Ulbrich MH, Isacoff EY (2007) Subunit counting in membrane-bound proteins. Nat Methods 4:319–321
23. Ren X, Li H, Clarke RW, Alves DA, Ying L, Klenerman D, Balasubramanian S (2006) Analysis of human telomerase activity and function by two color single molecule coincidence fluorescence spectroscopy. J Am Chem Soc 128:4992–5000
24. Ren X, Gavory G, Li H, Ying L, Klenerman D, Balasubramanian S (2003) Identification of a new RNA.RNA interaction site for human telomerase RNA (hTR): structural implications for hTR accumulation and a dyskeratosis congenita point mutation. Nucleic Acids Res 31:6509–6515
25. Tatavosian R, Zhen CY, Ren XJ (2015) Single-molecule fluorescence microscopy methods in chromatin biology. ACS Symp Ser 1215:129–136
26. Aggarwal V, Ha T (2016) Single-molecule fluorescence microscopy of native macromolecular complexes. Curr Opin Struct Biol 41:225–232
27. Murphy PJ, Cipriany BR, Wallin CB, Ju CY, Szeto K, Hagarman JA, Benitez JJ, Craighead HG, Soloway PD (2013) Single-molecule analysis of combinatorial epigenomic states in normal and tumor cells. Proc Natl Acad Sci U S A 110:7772–7777
28. Shema E, Jones D, Shoresh N, Donohue L, Ram O, Bernstein BE (2016) Single-molecule decoding of combinatorially modified nucleosomes. Science 352:717–721
29. Jain A, Liu R, Xiang YK, Ha T (2012) Single-molecule pull-down for studying protein interactions. Nat Protoc 7:445–452
30. Chung SH, Kennedy RA (1991) Forward-backward non-linear filtering technique for extracting small biological signals from noise. J Neurosci Methods 40:71–86

Chapter 11

Chromatin Immunoprecipitation of Skeletal Muscle Tissue

Amarjit Saini and Carl Johan Sundberg

Abstract

Chromatin immunoprecipitation (ChIP) is an invaluable method for studying interactions between histone proteins and genomic DNA regions and transcriptional regulation using antibodies to enrich genomic regions associated with these epitopes. Either to monitor the presence of histones with post-translational modifications at specific genomic locations or to measure transcription factor interactions with a candidate target gene, protein–DNA complexes are most commonly crosslinked using formaldehyde, which stabilizes these transient interactions. Chromatin is then fragmented to allow separation of genomic fragments bound by the histone or transcription factor of interest away from those that are unbound. Following immunoprecipitation, formaldehyde crosslinks are reversed and enriched DNA fragments are purified. While some investigators have successfully performed ChIP experiments from crosslinked skeletal muscle in cell culture, the process is relatively inefficient compared to whole tissue. This chapter provides protocols specifically designed for the crosslinking and immunoprecipitation of human skeletal muscle biopsy samples in preparation for chromatin immunoprecipitation-sequencing (ChIP-seq).

Key words Chromatin immunoprecipitation, ChIP, Muscle, Histone, ChIP-seq

1 Introduction

Genomic DNA in eukaryotic cells and tissues is packaged with special proteins termed histones to form protein/DNA complexes called chromatin. In a transcriptionally active state the DNA is accessible, representing an open conformation due to the relaxed state of histones whereas areas where DNA packaged into highly condensed forms are inaccessible to transcription factors or chromatin-associated proteins [1]. The dynamic states of chromatin structures tightly govern activation and function of the genome. Histone modifications, recognized as a post-translational modification (PTM), are critical for regulating chromatin structure and function, which can in turn affect many DNA-related processes, such as transcription [2, 3], recombination [4], DNA repair and replication [2, 5], and chromosomal organization [6]. Chromatin immunoprecipitation (ChIP) is a powerful experimental tool used for probing epigenetic changes in transcriptional regulation,

Neus Visa and Antonio Jordán-Pla (eds.), *Chromatin Immunoprecipitation: Methods and Protocols*,
Methods in Molecular Biology, vol. 1689, DOI 10.1007/978-1-4939-7380-4_11, © Springer Science+Business Media LLC 2018

mapping sites of DNA methylation, and interactions between histone proteins and a genomic DNA region [7].

There are two general methodologies for carrying out ChIP experiments. One (XChIP) uses formaldehyde, a reversible protein–DNA crosslinking agent that serves to fix or stabilize the protein–DNA interactions and allows for their isolation [7, 8]. Chromatin is then fragmented to allow separation of genomic fragments bound by the histone protein or transcription factor of interest away from those that are unbound [9]. Following antibody enrichment (immunoprecipitation) specific to a particular protein or histone modification, formaldehyde crosslinks are reversed and enriched DNA fragments are purified. ChIP-enriched genomic DNA fragments can be mapped to the genome and quantified using genomic microarrays (ChIP-chip), deep sequencing (ChIP-seq) or quantitative PCR. Typically, the crosslinking agent can be added directly to cell culture media to crosslink proteins to DNA in cultured cells [10]. Protein–DNA interactions in many tissues can be similarly crosslinked by immersion of minced tissue pieces in media containing the crosslinking agent [10, 11].

The other methodology uses native chromatin (NChIP) prepared by nuclease digestion of cell nuclei [12]. Native ChIP is mainly applicable for mapping the DNA target of histone modifiers. As histones wrap around DNA to form nucleosomes, they are naturally linked. Chromatin is sheared by micrococcal nuclease digestion, which cuts DNA at the length of the linker, leaving nucleosomes intact and providing DNA fragments ranging from one nucleosome (200 bp) to five nucleosomes (1000 bp) in length [12]. Thereafter, methods similar to XChIP are used for clearing the cell debris, immunoprecipitating the protein of interest, removing protein from the immunoprecipitated complex, and purifying and analyzing the complex-associated DNA [12]. The major advantage for NChIP is antibody specificity. This stems from the fact that most antibodies are raised against unfixed, synthetic peptidic antigens and that the epitopes they need to recognize in the XChIP may be disrupted or destroyed by formaldehyde crosslinking [12]. This is likely to explain the high efficiency of precipitated DNA compared with XChIP protocols [13]. In general, methods used for NChIP are not applicable to non-histone proteins (thus XChIP provides a broader spectrum of protein–DNA interactions to interrogate) and there is a danger of protein rearrangement during chromatin preparation and precipitation.

While some investigators have successfully performed ChIP experiments from crosslinked skeletal muscle myoblasts and myotubes [14], crosslinking of skeletal muscle tissue is relatively inefficient compared to most other tissues [10]. Skeletal muscle is composed of multiple muscle fibers (myofibers) that are responsible for producing force by contraction. The myofiber is a highly structured syncytium consisting of repeated arrays of contractile elements sustained by hundreds of peripherally located post-mitotic myonuclei. The myofiber structure may thus present a physical barrier that limits access of the crosslinking agent to the nuclei.

When material is limited, which is often the case with primary cell or tissue samples, smaller starting amounts of DNA have to be used. At present, ChIP-seq offers the best platform for investigating epigenetic changes in skeletal muscle. Only small amounts of DNA are required for library preparation ranging from 15 ng to as little as 50 pg using Helicos next-generation single-molecule sequencing platform. After strong enrichment of the chromatin, specifically bound by the protein under study, the material is ready to be sequenced. The DNA-binding proteins most frequently investigated in this way are transcription factors (e.g., p53 or NFκB), chromatin-modifying enzymes (e.g., p300, histone deacetylases), components of the basal transcriptional machinery (e.g., RNA polymerase II), and modified histones (e.g., H3K4me3) [15].

The current protocol supplements existing methods for XChIP (Agilent (https://www.agilent.com/cs/library/usermanuals/Public/G4481-90010_ChIP-on-chip_11.3.pdf) and Affymetrix (http://media.affymetrix.com/support/downloads/manuals/chromatin_immun_ChIP.pdf) and can be performed on whole fresh or frozen muscle biopsies. More and more researchers wish to investigate the molecular epigenetic mechanisms and patterns in relation to skeletal muscle phenotypes (muscle mass, mitochondrial and capillary abundance, substrate and metabolite transporters, etc.) in various states and conditions, including diabetes, cachexia, age and physical and nutritional status of the individuals under study. Skeletal muscle mass, quality and adaptability are fundamental in promoting muscle performance, maintaining metabolic function, supporting longevity, and health span. We believe that this protocol may help further investigations in this field.

2 Materials

As mentioned, the materials and methods have been adapted from published [7, 10] and commercial protocols from Agilent Inc. (Santa Clara, California) and Affymetrix Inc. (Santa Clara, California). Prepare all solutions using ultrapure water and analytical grade reagents. Prepare and store all reagents at room temperature (unless indicated otherwise). The protocol has been successfully developed for the extraction of crosslinked protein histone 3 trimethylated on lysine 27-DNA using rabbit anti-trimethyl-Histone H3 (Lys27) antibody on human skeletal muscle tissue and analyzed using ChIP-seq. However, the ChIP methods presented could be used to study various histone proteins, hormone receptors or transcription factor–DNA interactions utilizing specific antibodies to allow enrichment of the genomic regions.

2.1 Tissue Preparation and Crosslinking

1. Porcelain mortar and pestle.
2. Liquid nitrogen.
3. Spatula.
4. Cryo-gloves.
5. 1% formaldehyde in 50 mM HEPES-KOH pH 7.5, 100 mM NaCl, 1 mM EDTA pH 8.0 and 0.5 mM EGTA pH 8.0 supplemented with protease inhibitors.
6. Formaldehyde quenching solution of 2.5 M glycine.
7. Phosphate-buffered saline (PBS) supplemented with protease inhibitors.
8. 15 mL centrifuge tubes.
9. Tube rotator.

2.2 Tissue Lysis, Cell Harvesting, and Sonication

1. Dounce tissue grinder.
2. Benchtop centrifuge at 4 °C.
3. Lysis buffer 1 (tissue and cell lysis buffer): 50 mM HEPES-KOH pH 7.5, 140 mM NaCl, 1 mM EDTA pH 8.0, 10% glycerol, 0.5% Igepal (NP-40), 0.25% Triton X-100 and supplemented with protease inhibitors.
4. Lysis buffer 2 (nuclei wash buffer): 10 mM Tris–HCl pH 8.0, 200 mM NaCl, 1 mM EDTA pH 8.0, 0.5 mM EGTA pH 8.0 supplemented with protease inhibitors.
5. Lysis buffer 3 (nuclear lysis buffer): 10 mM Tris–HCl pH 8.0, 100 mM NaCl, 1 mM EDTA pH 8.0, 0.5 mM EGTA pH 8.0, 0.1% sodium deoxycholate, 0.5% N-lauroylsarcosine and supplemented with protease inhibitors.
6. PBS supplemented with protease inhibitors.
7. Sonicating water bath (e.g., Diagenode Bioruptor) or probe sonicator (with dry ice water bath).
8. Hemocytometer.
9. Trypan blue.
10. Light microscope.

2.3 Immunoprecipitation

1. Protein G magnetic beads and magnetic tube rack.
2. Block solution: 0.5% BSA (w/v) in PBS.
3. ChIP grade antibodies to target proteins (e.g., rabbit anti-H3K27me3); Negative control antibody/normal rabbit serum (recommended).
4. 1.5 mL microcentrifuge tubes.
5. Rotating platform/tube mixer or rocker at 4 °C.

6. RIPA ChIP wash buffer: 50 mM HEPES-KOH pH 7.6, 500 mM LiCl, 0.5 mM EDTA pH 8.0, 1% Igepal (NP-40), 0.7% sodium deoxycholate.
7. TE with 50 mM NaCl: 10 mM Tris–HCl pH 8.0, 1 mM EDTA pH 8.0 and 50 mM NaCl.
8. Elution buffer: 50 mM Tris–HCl pH 8.0, 10 mM EDTA pH 8.0, 1% SDS

2.4 DNA Isolation

1. TE: 10 mM Tris–HCl pH 8.0, 1 mM EDTA pH 8.0.
2. RNase A, 20 mg/mL (DNase free).
3. $CaCl_2$ solution: 300 mM $CaCl_2$ in 10 mM Tris–HCl pH 8.0.
4. Proteinase K, 20 mg/mL.
5. Phenol:chloroform:isoamyl alcohol (25:24:1).
6. Chloroform:isoamyl alcohol (24:1).
7. Glycogen, 20 mg/mL (Sigma) or suitable carrier for precipitation (i.e., sodium acetate, pH 5.2).
8. Ethanol 70% and absolute.
9. Fluorometer for DNA quantification (e.g., Qubit or similar).

3 Methods

3.1 Tissue Preparation and Crosslinking

1. Cryogenic grinding using a mortar and pestle. Grind frozen skeletal muscle biopsies/tissue samples ≥25 mg using a sterile cleaned mortar and pestle into small particles/powder and transfer the sample into a 15 mL centrifuge tube using a pre-chilled spatula (*see* **Notes 1** and **2**).
2. Add 5 mL of 1% formaldehyde solution and incubate on a rotary tube mixer for 20 min at room temperature (*see* **Note 3**).
3. Quench the formaldehyde crosslinking reaction by adding 1/20th volume of 2.5 M glycine solution directly to the 1% formaldehyde solution (to give a final concentration of 125 mM) and incubate on a rotary tube mixer for 5 min at room temperature.

3.2 Tissue Lysis, Cell Harvesting, and Sonication

1. Spin sample at 1350 × g for 5 min at 4 °C in a benchtop centrifuge with swinging bucket rotor.
2. Aspirate solution and resuspend pellet in 10 mL PBS supplemented with protease inhibitors and spin sample at 1350 × g for 5 min at 4 °C. Repeat once more.
3. Aspirate PBS and on ice/at 4 °C, resuspend pellet in 3 mL Lysis Buffer 1 (LB1) and transfer to a dounce tissue grinder (*see* **Note 4**) to facilitate tissue lysing. After douncing return sample to 15 mL centrifuge tube. Add a further 2 mL LB1 to dounce

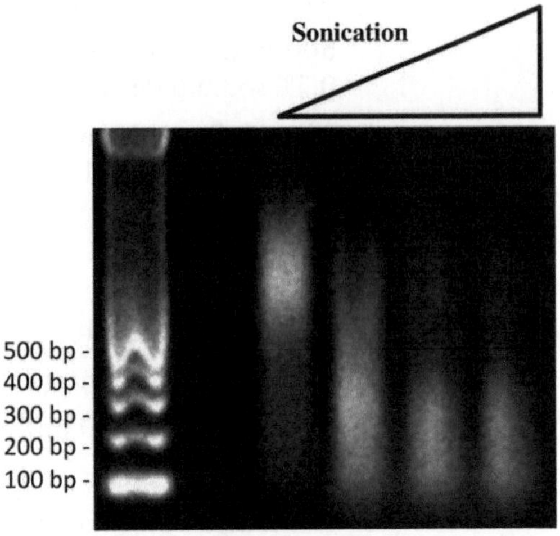

Fig. 1 Example of agarose gel electrophoresis on human skeletal muscle tissue to visualize/optimize chromatin shearing after sonication

tube to wash any remaining/residual sample from tube and add to 15 mL centrifuge tube for a total volume of 5 mL of LB1. Ensure steps are completed in 10 min at 4 °C/on ice. Spin sample at 1350 × *g* for 5 min at 4 °C.

4. Discard supernatant and resuspend pellet in 5 mL Lysis Buffer 2 (LB2) and rock gently at room temperature. Pellet nuclei by spinning at 1350 × *g* for 5 min at 4 °C (*see* **Note 5**).
5. Discard supernatant and resuspend pellet in 3 mL Lysis Buffer 3 (LB3).
6. Sonicate the suspension with a microtip attached to sonicator. Samples should be kept in an ice water bath during sonication (*see* **Notes 6** and 7).
7. Take 50 μL from the sample and assess the extent of sonication by electrophoresis on a 1% agarose gel after reversing formaldehyde crosslinks (follow Subheading 3.4, DNA isolation for procedure). A smear should be visible, with the majority of fragments between 250 bp and 1 kb (Fig. 1).
8. Add 300 μL of 10% Triton X-100 to the sonicated lysate and mix by pipetting up and down several times. Split into two 1.5 mL microfuge tubes. Centrifuge at 20,000 × *g* for 10 min at 4 °C in a microfuge to pellet debris.
9. Combine supernatants from the two 1.5 mL microfuge tube into a new 15 mL conical tube for immunoprecipitation.
10. Save 50 μL of lysate from the sample as total genomic input control and store at −20 °C (*see* **Note 8**).

3.3 Immunoprecipitation

The use of magnetic beads requires placing tubes in a magnetic stand and waiting for at least 1 min to immobilize the beads before washing beads or removing supernatants. This will prevent loss of beads or unintentional carryover.

1. Aliquot 100 μL of protein-G magnetic beads (*see* **Note 9**) per ChIP reaction to a 1.5 mL microfuge tube and wash three times with 1 mL block solution (PBS-BSA, 0.5%), collecting beads on a magnetic rack in between washes. Resuspend beads in 250 μL of PBS-BSA and add 10 μg of anti-H3K27me3 antibody (*see* **Note 10**) and incubate overnight on a rotating platform at 4 °C.
2. Next day, wash antibody–bead complexes three times with 1 mL PBS-BSA and resuspend in 100 μL block solution.
3. Add 100 μL of antibody–bead complexes to the 3 mL of sonicated lysates in 15 mL tubes and incubate overnight at 4 °C on a rotary mixer or rocker at low speed.
4. Working in a cold room at 4 °C transfer chromatin–antibody-–bead complexes to pre-chilled 1.5 mL centrifuge tubes, by sequentially adding 1 mL of the mixture to tubes on a magnetic rack and discarding the supernatant.
5. Wash beads seven times with 1 mL of ice-cold RIPA ChIP wash buffer, taking care to fully resuspending beads during washes by shaking/gently agitating.
6. Wash beads once with 1 mL ice-cold TE supplemented with 50 mM NaCl and discard the supernatant. Spin beads at 960 × *g* in a bench-top microfuge for 3 min at 4 °C and carefully discard the residual supernatant.
7. Working at room temperature elute ChIP material by adding 210 μL of ChIP elution buffer and incubating at 65 °C for 15 min. During elution, resuspend beads every 2 min by mixing briefly on a vortex mixer. Centrifuge samples at 16, 000 × *g* for 1 min at room temperature to pelet beads and immediately transfer the supernatant to a fresh 1.5 mL centrifuge tube.

3.4 DNA Isolation

1. Reverse protein–DNA crosslinks of ChIP samples by incubating at 65 °C overnight. For total genomic input control reserved after sonication add 3 volumes (150 μL) of elution buffer and mix. Reverse the crosslinks by incubating at 65 °C overnight.
2. Add 200 μL of TE to samples and genomic input control to dilute the SDS present in the elution buffer. Digest contaminating RNA by adding RNase A (20 μg/mL final concentration). Mix and incubate at 65 °C for 2 h.
3. Digest proteins by adding 7 μL $CaCl_2$ solution to each sample followed by 4 μL of 20 mg/mL proteinase K (20 μg/mL final concentration). Mix and incubate at 55 °C for 60 min.

4. Recover DNA by adding an equal volume (~400 μL) of phenol:chloroform:isoamyl alcohol, vortex vigorously for 1 min and centrifuge at 16,000 × *g* for 15 min at room temperature. If samples remain cloudy, repeat phenol:chloroform:isoamyl alcohol extraction one more time.
5. Carefully transfer the aqueous phase to a fresh 1.5 mL centrifuge tube, add an equal volume of chloroform:isoamyl 24:1 (~200 μL) to remove any residual phenol. Vortex mix the sample and centrifuge at 15,000 × *g* for 1 min.
6. Carefully transfer the aqueous phase to a fresh 1.5 mL centrifuge tube, add 30 μg glycogen, three volumes of absolute ethanol (~600 μL) and 200 mM NaCl. Cool the mixture for at least 30 min (to overnight) at −80 °C.
7. Pellet DNA by centrifugation at 20,000 × *g* for 30 min at 4 °C. Discard the supernatant.
8. Wash the DNA pellet with 500 μL of 70% ethanol and then centrifuge at 20,000 × *g* for 10 min at 4 °C to remove excess salts. Discard supernatant and air-dry under a lamp. Resuspend in 10 μL of 10 mM Tris pH 8.
9. Quantify and check the quality of the final immunoprecipitated DNA with fluorometric instruments. Use 10 ng for Illumina sequencing (*see* **Note 11**).

4 Notes

1. Grinding frozen samples with liquid nitrogen using a mortar and pestle is a widely used method. The mortar and pestle are cleaned and placed in a polystyrene tub or cooler where liquid nitrogen is poured or dispensed onto the mortar and pestle. Care is needed to avoid splattering liquid nitrogen when the mortar and pestle first start chilling. After several minutes the set will be cooled and a fog will usually settle over the apparatus. The sample may already be frozen or it can be snap frozen by dropping it into a beaker of liquid nitrogen (use plastic beakers) [16]. If the sample is taken from a −80 °C freezer, let it sit on the surface of the mortar to chill further. The relatively large mass of the mortar and pestle serve as a cold reservoir which helps to prevent sample thawing. It is important that frozen tissue samples do not reach high temperature (>4 °C), *work in cold is optimal* to prevent sample degradation by proteases. To grind, hold the pestle with a gloved hand (use a cryo-glove) and firmly press on the sample while twisting. The sample will typically shatter into small pieces. The fragmented pieces of the sample will continue to get smaller as the sample is ground using a circular motion with downward

pressure [16]. Once the grinding is completed, the ground sample must be tapped or scraped from the pestle. The sample must then be transferred into a receiving vessel using a pre-chilled spatula.

2. A significant problem in cryogenic grinding with mortar and pestle is that small samples (e.g., 10–20 mg) can be essentially lost when ground into the surface of the mortar. This makes sample recovery difficult and leads to poor yields [16]. Another major disadvantage of mortars and pestles is that the number of samples that can be processed is low. As the mortar and pestle may be in the −150 °C range following grinding, they must be warmed to room temperature (slowly) between uses and cleaned. Consequently, if many samples are processed daily, many sets are needed. Labs that process significant numbers of samples cryogenically must dedicate significant shelf space to the mortar and pestle sets [16]. Freeze drying is a method we have recently employed however this equipment may not be accessible to all groups. We have found that this improves yield. Following the procedure the muscle tissue is disrupted using a glass/dounce homogenizer on dry ice before proceeding to Subheading 3.1, **step 2**.
3. Ensure the fixation step is performed for the correct period of time using high quality, fresh formaldehyde. Very short or very long crosslinking time can lead to DNA loss and/or elevated background. Crosslinking can affect both efficiency of chromatin shearing and efficiency of specific antigen immunoprecipitation. Shorter crosslinking times (5–10 min) and/or lower formaldehyde concentrations (<1%, weight/volume) may improve shearing efficiency. However, for some proteins, especially those that do not directly bind DNA, this might reduce the efficiency of crosslinking and thus the yield of precipitated chromatin. It is possible to optimize the fixation step by testing different incubation times, such as 10, 20, and 30 min. Do not crosslink for longer than 30 min as such crosslinks cannot be efficiently sheared [17]. Two major problems concerning the subsequent immunoprecipitation step should be taken into account: (a) An excess of crosslinking can result in the loss of material or reduced antigen availability in chromatin, or both. (b) The relative sensitivity of the antigen epitopes to formaldehyde [18]. It is essential to perform the crosslinking step with care.
4. Though this device looks like a ground glass homogenizer, it relies on pushing a sample between the sides of the tube and the pestle. Shearing forces are created as the sample and liquid squeeze up and past the pestle. The Dounce homogenizer is most effective at lysing tissue culture cells and finely diced tissue in order to generate lysates where there are still intact subcellular particles [16]. Once the sample is placed in the

tube, the pestle is inserted, pressed down, and then lifted. This up and down motion is repeated, causing the sample to be sheared repeatedly. The shearing force can be controlled to an extent by using different pestles with different diameters. The larger diameter pestle has tighter fitting and creates a greater shear, while the opposite is true for the smaller pestle.

5. It is possible to check the release and integrity of nuclei under a light microscope after staining a small aliquot (5–10 μL) with an equal volume of trypan blue dye. Nuclei will stain blue while any remaining unlysed tissue will clear the dye and appear unstained. If lysis is insufficient, resuspend pellet in 5 mL Lysis Buffer 1 (LB1) and rock at 4 °C for 10 min before proceeding to Subheading 3.2, **step 4**.

6. Using a Branson Sonicator set power to 6 and duty cycle at 20%. Sonicate 16 cycles of 30 s ON and 60 s OFF to decrease foaming. Sonication settings will need to be optimized (*see* **Note** 7) that result in sheared DNA that ranges from 100 to 600 bp in size.

   We have found a dry ice water bath to be most effective using a single-probe sonicator. Sonication generates a tremendous amount of heat, which can denature many proteins. This can be offset by short bursts coupled with incubations on ice/dry ice.

7. It is important to optimize sonication conditions since this step defines the resolution of ChIP. It is advisable to test a range of conditions including length of pulse, number of pulses, and amplitude of sonication. The efficiency of sonication should be assessed by resolving a sample of total chromatin after sonication and de-crosslinking using agarose gel electrophoresis (Fig. 1). Overall, the goal is to obtain consistent results using sonication. Slight differences in volume, placement of the probe tip in vessel, foaming, and other subtle changes may lead to different shearing results.

8. It is important in a ChIP-seq experiment to run a control using "input DNA" (non-ChIP genomic DNA) so that sequencing biases can be identified and adjusted for.

9. Bind the antibody to the beads in a cold room or on ice. The exact type of magnetic bead (Protein A, Protein G) depends on the antibody being used.

10. Depending on biopsy availability a parallel sample with Negative Control antibody (normal rabbit serum) is strongly recommended. Normal rabbit serum can act as negative control to help differentiate nonspecific background signal from specific antibody signal because there is no relevant specificity to a target antigen.

11. We analyze the quality of ChIP DNA using Tapestation High sensitivity D1000 screentape (Agilent Technologies, Santa Clara, CA) and quantify using Qubit dsDNA HS assay kit with a Qubit Fluorometer (Life Technologies, Carlsbad, CA). In our experiments, 10 ng of ChIP DNA was used as input in NEBNext ChIP-Seq library prep (NEB). Four sample libraries were multiplexed and run at 50 cycles single reads per lane on Illumina HiSeq 2000. The instruction manual for sample preparation and preparation of expression libraries guidebook from New England Biolabs is free to download (https://www.neb.com/~/media/Catalog/All-Products/7DEB5A53AA7D42869B1981EF97F90F6F/Datacards%20or%20Manuals/manualE6240.pdf).

Acknowledgments

We thank Patrick Muller at BEA—the core facility for Bioinformatics and Expression Analysis—Karolinska Institutet, Huddinge for quantifying ChIP enrichments and mapping ChIP-enriched genomic DNA fragments to the genome using ChIP-seq as well as help with editing the protocol.We thank Fiona Myers at the School of Biological Sciences, University of Portsmouth for her scientific advice and extensive knowledge in XChIP, NChIP, ChIP-seq and help with editing the protocol.

References

1. Saini A, Mastana S, Myers F et al (2013) 'From death, lead me to immortality'–mantra of ageing skeletal muscle. Curr Genomics 14:256–267
2. Consortium EP, Birney E, Stamatoyannopoulos JA et al (2007) Identification and analysis of functional elements in 1% of the human genome by the ENCODE pilot project. Nature 447:799–816
3. Koch CM, Andrews RM, Flicek P et al (2007) The landscape of histone modifications across 1% of the human genome in five human cell lines. Genome Res 17:691–707
4. Krivtsov AV, Feng Z, Lemieux ME et al (2008) H3K79 methylation profiles define murine and human MLL-AF4 leukemias. Cancer Cell 14:355–368
5. Groth A, Rocha W, Verreault A et al (2007) Chromatin challenges during DNA replication and repair. Cell 128:721–733
6. Jenuwein T, Allis CD (2001) Translating the histone code. Science 293:1074–1080
7. Massie CE, Mills IG (2011) Global identification of androgen response elements. Methods Mol Biol 776:255–273
8. Orlando V (2000) Mapping chromosomal proteins in vivo by formaldehyde-crosslinked-chromatin immunoprecipitation. Trends Biochem Sci 25:99–104
9. Carey MF, Peterson CL, Smale ST (2009) Chromatin immunoprecipitation (ChIP). Cold Spring Harb Protoc 2009:pdbprot5279
10. Ohkawa Y, Mallappa C, Vallaster CS et al (2012) Isolation of nuclei from skeletal muscle satellite cells and myofibers for use in chromatin immunoprecipitation assays. Methods Mol Biol 798:517–530
11. Wells J, Farnham PJ (2002) Characterizing transcription factor binding sites using formaldehyde crosslinking and immunoprecipitation. Methods 26:48–56
12. Thorne AW, Myers FA, Hebbes TR (2004) Native chromatin immunoprecipitation. Methods Mol Biol 287:21–44

13. Das PM, Ramachandran K, vanWert J et al (2004) Chromatin immunoprecipitation assay. BioTechniques 37:961–969
14. Mejat A, Ramond F, Bassel-Duby R et al (2005) Histone deacetylase 9 couples neuronal activity to muscle chromatin acetylation and gene expression. Nat Neurosci 8:313–321
15. Liu ET, Pott S, Huss M (2010) Q&A: ChIP-seq technologies and the study of gene regulation. BMC Biol 8:56
16. Burden DW (2012) Guide to the disruption of biological samples. Random Primers 12:1–25
17. Tian B, Yang J, Brasier AR (2012) Two-step cross-linking for analysis of protein-chromatin interactions. Methods Mol Biol 809:105–120
18. Crane-Robinson C, Myers FA, Hebbes TR et al (1999) Chromatin immunoprecipitation assays in acetylation mapping of higher eukaryotes. Methods Enzymol 304:533–547

Chapter 12

Chromatin Immunoprecipitation Assay in the Hyperthermoacidophilic Crenarchaeon, *Sulfolobus acidocaldarius*

Kun Wang and Ann-Christin Lindås

Abstract

Chromatin immunoprecipitation (ChIP) is a powerful method used for identifying genome-wide DNA--protein interactions in vivo. A large number of essential intracellular processes such as DNA replication, transcription regulation, chromatin stability, and others are all dependent on protein interactions with DNA. The DNA fragments enriched from the ChIP assay are analyzed by downstream applications, for example, microarray hybridization (ChIP-chip), quantitative PCR (ChIP-qPCR), or deep sequencing (ChIP-seq). This chapter presents a stepwise protocol for ChIP performed in hyperthermophilic archaea that we have successfully used in the hyperthermoacidophilic crenarchaeon *Sulfolobus acidocaldarius*.

Key words Chromatin immunoprecipitation, ChIP, Archaea, *Sulfolobus*, Hyperthermophile

1 Introduction

Many of the fundamental and vital events occurring within a cell are completely dependent on DNA–protein interactions. Such events are DNA replication, transcription, chromatin formation and stabilization by histones in eukaryotes, DNA repair and recombination, DNA methylation, etc. The development of chromatin immunoprecipitation (ChIP) has enabled the studies of these DNA–protein interactions at a genome-wide scale in vivo. Starting in 1984, the first ChIP assays were performed in bacteria and shortly followed by fruit fly to study the interaction between DNA polymerase and DNA [1, 2]. Since then, extensive improvements have been made and ChIP is now used within a number of research areas such as transcription regulation, histone–DNA interactions and histone modifications, development, homeostasis regulation in diseases, etc., as well as in a number of different species ([3] and references therein). The method is based on the following main steps: (1) cross-linking of the DNA and proteins in the cells by adding a cross-

Neus Visa and Antonio Jordán-Pla (eds.), *Chromatin Immunoprecipitation: Methods and Protocols*, Methods in Molecular Biology, vol. 1689, DOI 10.1007/978-1-4939-7380-4_12, linking agent such as formaldehyde into the cell culture or to the tissue, (2) disruption of the cells and fragmentation of the DNA, cross-linked with proteins, by digestion with enzymes or by sonication, (3) addition of antibodies specific for the protein of interest followed by isolation of the antibody–protein–DNA complex, and finally (4) reverse cross-linking and purification of the enriched DNA fragments. The DNA fragments are, thereafter, analyzed by use of downstream applications, e.g., microarray analysis (ChIP-chip), qPCR (ChIP-qPCR), or deep sequencing (ChIP-seq).

As late as 1990, by the work of Carl Woese and colleagues, Archaea was recognized as a separate domain beside Bacteria and Eukarya [4]. The achievements within the research field of archaea are therefore relatively recent. As for the mechanisms of transcription regulation in archaea, little is known. Archaea seem to have a reduced number of transcription factors [5], compared with bacteria, and for the majority of them, the DNA target sequences are unknown. In our research, we use ChIP combined with deep sequencing (ChIP-seq) to identify DNA target sequences for different transcription factors in the archaeal genus *Sulfolobus*, and subsequently identify potential target genes that are regulated by the transcription factor of interest. Here, we provide a thorough description of how to perform ChIP using the hyperthermophilic model organism *Sulfolobus acidocaldarius*.

2 Materials

2.1 Cell Culture

1. Long neck flasks (*see* **Note 1**).
2. Shaking bath running at 80 °C (*see* **Note 2**).
3. Spectrophotometer.
4. Strain *S. acidocaldarius* DSM639.
5. Trace mineral solution: 9.0 g/L $Na_2B_4O_7 \cdot 10H_2O$, 3.6 g/L $MnCl_2 \cdot 4H_2O$, 0.44 g/L $ZnSO_4 \cdot 7H_2O$, 0.10 g/L $CuCl_2 \cdot 2H_2O$, 0.06 g/L $Na_2MoO_4 \cdot 2H_2O$, 0.079 g/L $VOSO_4 \cdot 5H_2O$, 0.10 g/L $CoSO_4 \cdot 7H_2O$. Sterile filter through a 0.22 μm filter.
6. 2×Allen culture medium: 2.60 g/L $(NH_4)_2SO_4$, 0.56 g/L KH_2PO_4, 0.5 g/L $MgSO_4 \cdot 7H_2O$, 0.14 g/L $CaCl_2 \cdot 2H_2O$, 1 mL/L 148 mM $FeCl_3 \cdot 6H_2O$ solution, 1 mL/L Trace Mineral Solution, adjust to pH 2.0 using H_2SO_4 and filter-sterilize through a 0.22 μm filter (*see* **Note 3**).
7. 5% (w/v) tryptone. Sterilize by autoclaving.
8. 1×Allen culture medium (1 L): 500 mL 2×Allen, 40 mL 5% tryptone, and 460 mL sterile H_2O.

2.2 In Vivo Cross-Linking and Sonication

1. Probe sonicator (Soniprep 150 with 3 mm diameter probe is used in this protocol).
2. Centrifuges (Sorvall RC 5C plus floor centrifuge and table-top centrifuge).
3. Shaking bath running at 80 °C (*see* **Note 2**).
4. Shaker at room temperature.
5. 36.5% formaldehyde.
6. 2.5 M glycine in H_2O, sterilized using a 0.22 μm filter.
7. Phosphate Buffer Saline (PBS) pH 7.4: 8.0 g/L NaCl, 0.2 g/L KCl, 1.8 g/L $Na_2HPO_4{\cdot}2H_2O$, 0.24 g/L KH_2PO_4. Sterilize by autoclaving.
8. 10% (w/v) sodium deoxycholate in H_2O. Sterilize using a 0.22 μm filter.
9. IP buffer: 50 mM Hepes-KOH, pH 7.5, 150 mM NaCl, 1 mM EDTA, 1% (w/v) Triton X-100, 0.1% (w/v) sodium deoxycholate, 0.1% SDS (w/v), cOmpete protease inhibitor cocktail according to the manufacturer's instructions (Roche Applied Science).

2.3 DNA Extraction Using Phenol/Chloroform

1. Vortex.
2. Table-top centrifuge.
3. 10 mg/mL RNase A.
4. 10 mg/mL Protease K.
5. Phenol:chloroform:isoamylalcohol (25:24:1).
6. Chloroform:isoamylalcohol (24:1).
7. 3 M sodium acetate (NaAc), pH 5.2.
8. 100% ethanol.
9. DNase-free water.

2.4 Agarose Gel Electrophoresis

1. Equipment for agarose gel electrophoresis.
2. Agarose.
3. 1× Tris-acetate-EDTA (TAE) buffer: 40 mM Tris–HCl, pH 7.6, 20 mM acetic acid, 1 mM EDTA.
4. Staining for DNA (we use ethidium bromide).

2.5 Immunoprecipitation

1. Vortex.
2. Rotator for 1.5 mL tubes.
3. Dynabeads® M-280 Sheep Anti-Rabbit IgG magnetic beads (Life Technologies).
4. Magnetic rack for 1.5 mL tubes.
5. Polyclonal rabbit antibody specific for target protein (*see* **Note 4**).

6. Rabbit pre-immune serum (*see* **Note 4**).
7. Phosphate Buffer Saline (PBS), pH 7.4: 8.0 g/L NaCl, 0.2 g/L KCl, 1.8 g/L $Na_2HPO_4 \cdot 2H_2O$, 0.24 g/L KH_2PO_4. Sterilize by autoclaving.
8. IP wash buffer 2: same as IP buffer (Subheading 2.2) but with 500 mM NaCl.
9. IP wash buffer 3: 10 mM Tris–HCl, pH 8.0, 250 mM LiCl, 1 mM EDTA, 0.5% (w/v) NP-40, 0.5% (w/v) sodium deoxycholate.
10. Tris–EDTA (TE) buffer: 10 mM Tris–HCl, pH 8.0, 1 mM M EDTA. Sterilize by autoclaving.

2.6 Post-immunoprecipitation

1. IPure kit (Diagenode).
2. 100% isopropanol.
3. 3 M sodium acetate (NaAc), pH 7.0.
4. 20 mg/mL glycogen.
5. 100% ethanol.
6. 70% ethanol.
7. DNase-free water.

3 Methods

3.1 Culture of *Sulfolobus acidocaldarius* DSM639

1. Start *Sulfolobus acidocaldarius* DSM639 culture by adding 50 μL from a glycerol stock to a 50 mL long neck flask containing 20 mL 1×Allen culture medium. Grow the culture at 80 °C with shaking at 140 rpm and monitor culture growth by measuring the optical density at 600 nm.
2. When OD_{600nm} of the culture reaches 0.4 (mid-exponential phase), transfer 1 mL culture to a new 500 mL long neck flask containing pre-heated 200 mL 1×Allen culture medium. Continue culturing at 80 °C with shaking at 140 rpm.

3.2 In Vivo Cross-Linking

1. Measure OD_{600nm} regularly and when the culture reaches the desired cell density, transfer the culture to a shaker at room temperature and immediately add 5.4 mL of 36.5% formaldehyde (final concentration 1% v/v) slowly to the culture while shaking (*see* **Note 5**). Continue shaking for 5 min at 40 rpm.
2. Thereafter, add 10 mL 2.5 M glycine to a final concentration of 125 mM for quenching the cross-link. Continue shaking for 5 min at 40 rpm.
3. Collect the fixed cells by centrifugation at 8000 × *g* for 10 min at 4 °C.

4. Resuspend the pellet in 40 mL cold PBS and centrifuge at 8000 × *g* for 10 min at 4 °C (repeat this wash step three times in total).
5. The pellet can be stored at −80 °C or continue to sonication.

3.3 Sonication

1. Resuspend pellet from **step 4** in Subheading 3.2 in 3 mL IP buffer and transfer to a 15 mL conical tube.
2. Sonicate (on ice) to have an average DNA fragment size of about 300 bp. Check DNA size distribution by resolving 10 μL sonicated DNA on a 1.2% agarose gel (*see* **Notes 6** and **7**).
3. Divide the sonicated DNA mixture into two 2 mL tubes and centrifuge at 16,100 × *g* for 20 min at 4 °C.
4. Transfer and combine cleared supernatants into a new 15 mL tube. Save 100 μL cleared supernatant in a 1.5 mL tube for use as input control sample and store at −80 °C.

3.4 DNA Isolation for Checking Fragment Size Using Phenol/Chloroform Extraction

1. Take 50 μL sonicated sample and reverse the cross-link by incubating for 10 min at 90 °C.
2. When the tube has cooled down, add 1 μL of 10 mg/mL RNase A and 1 μL of 10 mg/mL Protease K and incubate for 20 min at 37 °C.
3. Add 50 μL phenol:chloroform:isoamylalcohol (25:24:1) to the tube, vortex and centrifuge at 16,100 × *g* for 5 min at room temperature. Transfer the upper phase to a new tube.
4. Add 50 μL chloroform:isoamylalcohol (24:1) to the new tube, vortex and centrifuge at 16,100 × *g* for 5 min at room temperature. Transfer the upper phase to a new tube.
5. Add 4 μL of 3 M NaAc and 500 μL of cold 100% ethanol. Invert the tube to mix well and place the tube at −20 °C for 2 h.
6. Centrifuge the tubes at 16,100 × *g* for 15 min at 4 °C. Carefully pipette off the supernatant.
7. Add 500 μL cold 70% ethanol. Centrifuge as in **step 6** and discard the supernatant.
8. Air-dry the pellet (do not completely dry the pellet) and dissolve in 10 μL DNase-free water.
9. Place the DNA on ice or store at −20 °C for further use.
10. Resolve the DNA sample on a 1.2% agarose gel to check DNA fragment size.

3.5 Agarose Gel Electrophoresis

1. Mix agarose with 1× TAE buffer, 1.2% (w/v) agarose. Melt the agarose using a microwave oven.
2. Chill the agarose solution to approx. 60 °C, add 1 μL/mL ethidium bromide, pour the agarose into the gel tray and put in the comb. Let the gel polymerize (20–30 min).

3. Prepare the samples by mixing 1:5 (v/v) DNA sample and 6× loading dye.
4. Load the samples and an appropriate molecular weight marker, and run the electrophoresis in 1× TAE buffer during 45–50 min at 80 V (time and voltage can vary). Visualize the DNA fragments using a UV-table and a suitable camera device.

3.6 Preparation of Antibody–Bead Complexes

1. Resuspend the Dynabeads® M-280 Sheep Anti-Rabbit IgG magnetic beads thoroughly in the vial by vortexing at least 30 s (*see* **Note 8**).
2. Transfer 80 μL of resuspended beads into two 1.5 mL tubes, respectively. Place the tubes in the magnetic rack for 1 min and pipette off the supernatant. Avoid touching the beads on the wall.
3. Remove the tubes from the magnetic rack and resuspend the pellet in 200 μL of PBS, reinsert the tubes in the magnetic rack and carefully remove the supernatant. Repeat this wash step three times in total.
4. Thereafter resuspend the beads in 80 μL PBS. Add 7 μg specific antibody to one tube with washed beads (IP), and add the same amount of pre-immune serum to the other tube as a mock control. If there is no pre-immune serum available, add the same volume of PBS to the washed beads to have a no-antibody mock control.
5. Incubate the two tubes on a rotator for 90 min at 4 °C.
6. Place the tubes in the magnetic rack for 2 min and pipette off the supernatant.
7. Remove the tubes from the magnetic rack, add 200 μL PBS respectively and resuspend the beads. Replace the tubes in the magnetic rack and remove the supernatants. Repeat the wash with PBS three times in total. The beads are ready for the immunoprecipitation.

3.7 Immunoprecipitation

1. Add 1 mL cleared cell lysate from Subheading 3.3 into two tubes containing the Antibody-Bead complexes prepared in Subheading 3.6.
2. Resuspend beads and incubate at 4 °C on a rotator overnight. Shorter incubation time is possible if the target protein is in high cellular concentration.
3. Place the tubes in the magnetic rack for 2 min and pipette off the supernatant.
4. Wash the beads twice using 100 μL of IP buffer each time and change buffers by the use of the magnetic rack.
5. Wash the beads once using 100 μL of IP wash buffer 2, put the tubes in the magnetic rack and pipette off the buffer.

6. Wash the beads once using 100 μL of IP wash buffer 3, put the tubes in the magnetic rack and pipette off the buffer.
7. Wash the beads once using 100 μL TE buffer, put the tubes in the magnetic rack and pipette off the buffer.

3.8 DNA Extraction and Purification

Immunoprecipitated DNA is extracted by IPure kit (Diagenode) according to the manufacturer's instructions (*see* **Note 9**).

4 Notes

1. Preferably use long neck flask for cultivation to minimize evaporation of the culture media. Seal the flask with aluminum foil.
2. For cultivation we use shaking incubators filled with mineral oil (Rotitherm® K+H, Carl Roth) instead of water to avoid evaporation. To avoid large temperature changes of the cultures, keep the flask in the incubator as far as possible when sampling and handling cultures. Shaking incubators with air are not recommended since just opening them causes rapid decrease in temperature.
3. When filter sterilizing larger volumes, use preferably the Filtropur 0.22 μm device from Sarstedt.
4. The polyclonal antibodies were produced by Innovagen, Lund, Sweden. One rabbit/antibody was immunized with the specific protein. A serum sample was taken from the same rabbit before immunization and used as control (pre-immune serum). The antibodies were purified in two steps: (a) IgG affinity purified and thereafter (b) affinity purified against the specific antigen. It is important that the antibodies used have high affinity to the specific protein.
5. Add formaldehyde directly to the culture at high temperature without cooling it down first [6, 7]. As formaldehyde is highly volatile and very toxic upon inhalation, work in the hood and seal the flask during fixation.
6. Sonication should be optimized depending on the downstream analysis. For deep DNA sequencing, it is recommended to have DNA fragments' average size around 300 bp. For quantitative real-time PCR or microarray, longer fragments are preferred (0.5–1 kb). The duration of the sonication varies depending on the type of strain, cell density, duration of formaldehyde fixation, and the brand of the sonicator.
7. To more clearly see the DNA size distribution on an agarose gel, we recommend to purify a small aliquot of the sample after the sonication step using phenol/chloroform extraction (Subheading 3.4).

8. Dynabeads® (Life Technologies) have several variations to choose from. As our specific antibody was raised in rabbit, Dynabeads® M-280 Sheep Anti-Rabbit IgG magnetic beads were chosen in this protocol. For mouse antibodies, Dynabeads® M-280 Sheep Anti-Mouse IgG is available. For other antibodies, beads coupled with protein A or protein G can be used based on their affinity.
9. If at least one DNA target region is known for your protein of interest, real-time PCR can be used to determine the relative enrichment of DNA fragment from the ChIP experiment. As reference region, you select a region that your protein does not bind to. After amplifying both regions from your purified enriched DNA sample and from the input (Subheading 3.3), the relative enrichment can be calculated by using the $2^{-\Delta\Delta CT}$ Method [8] as follows:

$$\Delta Ct \text{ (input sample)} = \text{Ct (target region)} - \text{Ct (reference region)}.$$

Ct is the "threshold cycle" and indicates the number of PCR cycles needed to obtain a significant fluorescence signal above the background. The Ct value is given by the PCR machine software.

$$\Delta Ct \text{ (enriched DNA sample)} = \text{Ct (target region)} - \text{Ct (reference region)}.$$

$$\Delta\Delta \text{Ct} = \Delta Ct \text{ (enriched DNA sample)} - \Delta Ct \text{ (input sample)}.$$

Fold-ratio enrichment enriched DNA sample/input sample = $2^{-\Delta\Delta \text{Ct}}$.

References

1. Gilmour DS, Lis JT (1984) Detecting protein-DNA interactions *in vivo*: distribution of RNA polymerase on specific bacterial genes. Proc Natl Acad Sci U S A 81:4275–4279
2. Gilmour DS, Lis JT (1985) *In vivo* interactions of RNA polymerase II with genes of *Drosophila melanogaster*. Mol Cell Biol 5:2009–2018
3. Kuo MH, Allis DC (1999) *In vivo* cross-linking and immunoprecipitation for studying dynamic protein:DNA associations in a chromatin environment. Methods 19:425–433
4. Woese CR, Kandler O, Wheelis ML (1990) Towards a natural system of organisms: proposal for the domains Archaea, Bacteria, and Eucarya. Proc Natl Acad Sci U S A 87:4576–4579
5. Pérez-Rueda E, Janga SC (2010) Identification and genomic analysis of transcription factors in archaeal genomes exemplifies their functional architecture and evolutionary origin. Mol Biol Evol 27:1449–1459
6. Liu H, Wang K, Lindås AC, Peeters E (2016) The genome-scale DNA-binding profile of BarR, a β-alanine responsive transcription factor in the archaeon *Sulfolobus acidocaldarius*. BMC Genomics 17:569
7. Reichelt R, Gindner A, Thomm M, Hausner W (2016) Genome-wide binding analysis of the transcriptional regulator TrmBL1 in *Pyrococcus furiosus*. BMC Genomics 8:40
8. Livak KJ, Schmittgen TD (2001) Analysis of relative gene expression data using real-time quantitative PCR and the $2^{-\Delta\Delta Ct}$ method. Methods 25:402–408

Chapter 13

Using Intra-ChIP to Measure Protein–DNA Interactions in Intracellular Pathogens

Brett R. Hanson and Ming Tan

Abstract

Chromatin immunoprecipitation is used to measure the binding of transcription factors to target DNA sequences in order to better understand transcriptional regulation. Here, we describe a process to analyze bacterial transcription factor binding in the context of an infected eukaryotic host cell. Using this approach, we measured the binding kinetics of three *Chlamydia trachomatis* transcription factors within infected cells, and demonstrated temporal changes in binding.

Key words Chromatin immunoprecipitation, Intracellular pathogen, Transcriptional regulation, *Chlamydia*

1 Introduction

Chromatin immunoprecipitation (ChIP) utilizes antibody specificity to selectively capture protein–DNA complexes for analysis [1]. Cells and tissues are first treated with a crosslinking agent, such as formaldehyde, to covalently link proteins that are bound to DNA. Specific antibodies are then used to preferentially isolate a particular protein cross-linked to its DNA targets. Bound DNA is subsequently dissociated from the protein by reversing the crosslinks, and can be analyzed using a variety of techniques.

ChIP is a critical component of modern molecular biology, but has only recently been used to probe gene regulation in microbial pathogens within an infected cell. Here, we demonstrate a ChIP method, called Intra-ChIP [2, 3], to measure binding of a bacterial transcription factor to its target genes in the context of an intracellular infection. We have used *Chlamydia trachomatis* as a representative intracellular pathogen, and describe our approach to detect binding of multiple chlamydial transcription factors to their regulatory DNA target sequences. *C. trachomatis* utilizes an unusual developmental cycle for intracellular growth [4], which involves temporal control of transcription [5]. We used the Intra-ChIP

Neus Visa and Antonio Jordán-Pla (eds.), *Chromatin Immunoprecipitation: Methods and Protocols*,
Methods in Molecular Biology, vol. 1689, DOI 10.1007/978-1-4939-7380-4_13, © Springer Science+Business Media LLC 2018

approach to measure intracellular binding of the chlamydial transcription factors EUO, HrcA, and Sigma-28.

We were able to determine that EUO binds its target DNA early in the chlamydial developmental cycle, and that binding is lost over time (Fig. 1a). We were also able to demonstrate that HrcA binds its target DNA at middle and late times of the chlamydial developmental cycle, and that HrcA binding is minimal at earlier times (Fig. 1b). Finally, we observed that Sigma-28 binds to its target DNA exclusively at late times in the developmental cycle (Fig. 1c). Thus, we have shown that transcription factor binding can be measured in an intracellular pathogen. Furthermore, temporal changes in transcriptional regulation can be monitored over time. In addition, the method described is broadly applicable to other intracellular microbes.

2 Materials

Prepare solutions using sterile ultrapure water and analytical grade reagents. Prepare and store all materials at room temperature, unless otherwise specified. Follow all relevant regulations and guidelines to dispose of waste appropriately.

2.1 Cell Culture

1. Cell Culture Medium: RPMI with 25 mM HEPES and 5% FBS (heat-inactivated). Prepare by combining 925 mL of RPMI 1640 with 25 mL of 1 M HEPES and 50 mL FBS. Filter through a 0.22 μm Corning filter. Store at 4 °C.
2. 500 mL spinner flask.
3. L929 mouse fibroblast cells (ATCC).
4. *Chlamydia trachomatis* Serovar L2, strain 434/Bu (ATCC).

2.2 Crosslinking

1. Crosslinking Solution: 1% formaldehyde in RPMI 1640 lacking HEPES and FBS. Prepared by adding an appropriate volume of fresh 37% formaldehyde directly to RPMI 1640.
2. Quenching Solution: 2.5 M glycine pH 7.5.

2.3 Lysate Preparation

1. Lysis Buffer: 25 mM Tris–HCl pH 7.5, 150 mM NaCl, 0.5% sodium deoxycholate, 1% NP-40, 0.1% SDS, protease inhibitor cocktail.
2. Branson Digital Sonifier.

2.4 Immunoprecipitation

1. GE Healthcare Life Sciences Protein G Sepharose, stored at 4 °C.
2. Antibody: Antigen-specific Serum and Control Serum, stored at −20 °C.

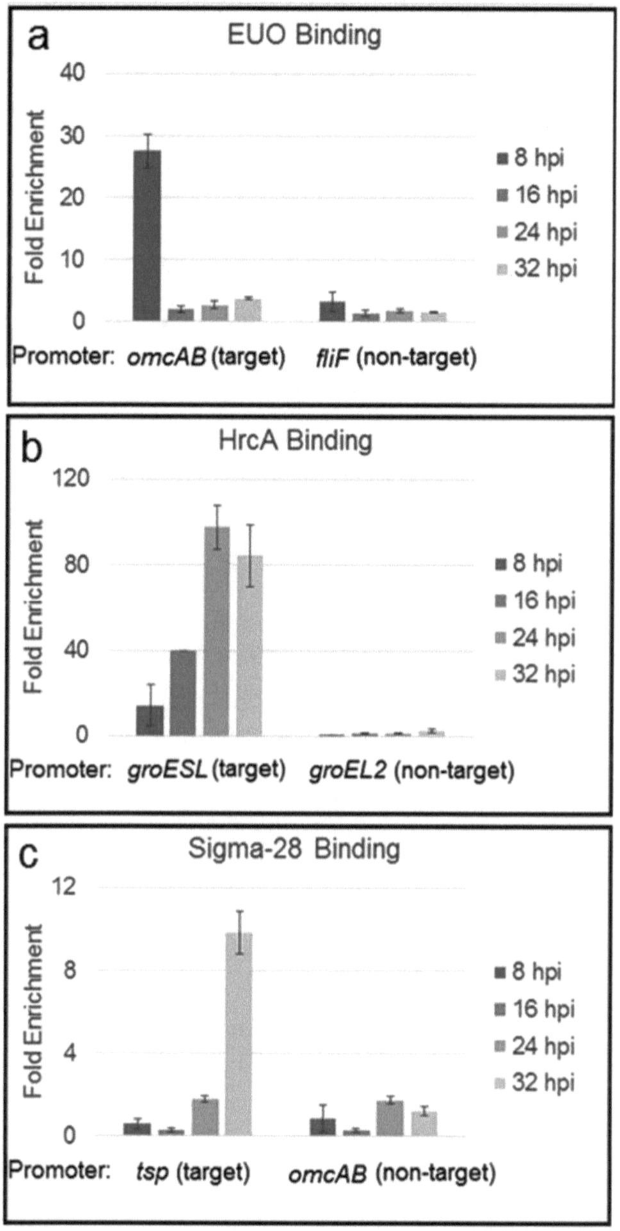

Fig. 1 Measurement of chlamydial transcription factor binding to target promoters using Intra-ChIP. The transcription factors EUO (**a**), HrcA (**b**), and Sigma-28 (**c**) were immunoprecipitated at multiple times after infection (hours post-infection, hpi) to measure binding to target or non-target DNA sequences. Binding is expressed as Fold Enrichment relative to immunoprecipitation with Control Serum

3. Blocking buffer: 5% BSA with 200 μg/mL Salmon sperm DNA, stored at 4 °C.
4. Phosphate Buffered Saline (PBS).

5. Wash buffer: 40 mM HEPES, 4 mM $MgCl_2$, 70 mM KCl, 7.5% glycerol.
6. Elution buffer: 0.1 M glycine pH 2.5.
7. Neutralization buffer: 1 M Tris–HCl pH 8.5.
8. Machery-Nagel Nucleospin Gel and PCR Clean-up Kit.

2.5 Quantitative Real-Time PCR Analysis

1. Bio-Rad SYBR Green master mix, stored at −20 °C.
2. Bio-Rad iCycler Real-Time PCR Detection System.
3. Primers to amplify Target DNA and Non-target DNA.

3 Methods

3.1 Cell Culture and Infection

1. Grow L929 cells in Cell Culture Medium at 37 °C in 5% CO_2. To produce sufficient cells for this protocol, we grow them in suspension culture in a 500 mL spinner flask at a density of 6×10^5 cells/mL (*see* **Note 1**).
2. Infect cells by directly adding *C. trachomatis* at a multiplicity of infection (MOI) of 3 to L929 suspension culture (*see* **Note 2**).
3. At desired time point post-infection, collect 6×10^7 infected cells (~100 mL) with centrifugation at 2500 × *g* for 10 min.

3.2 Crosslinking

1. Resuspend cells in 30 mL Crosslinking Solution and incubate gently, rocking for 30 min at room temperature (*see* **Note 3**).
2. Add 3.3 mL Quenching Solution to crosslinked cells and incubate gently, rocking for 15 min at room temperature (*see* **Note 4**).
3. Pellet crosslinked cells with centrifugation at 2500 × *g* for 10 min.
4. Aspirate supernatant and store infected cell pellets at −80 °C, or proceed directly to Lysate Preparation.

3.3 Lysate Preparation

1. Resuspend crosslinked infected cells in 4 mL Lysis Buffer in a 15 mL conical tube.
2. Sonicate on ice with an output setting of 22% for a total of 2 min in 15 s bursts. Sonication conditions should result in fragmentation of DNA to a size range between 300 and 1200 bp (*see* **Note 5**).
3. Centrifuge at 10,000 × *g* for 30 min to clarify lysate.
4. Carefully remove lysate supernatant without dislodging pellet.
5. Pre-clear crosslinked lysate by adding 50 μL (bed volume) of Protein G Sepharose and incubating with gentle rocking at 4 °C for 2 h (*see* **Note 6**).

6. Pellet Protein G Sepharose by centrifuging at 10,000 × *g* for 5 min.
7. Carefully remove lysate without disturbing Protein G Sepharose pellet.
8. Separate lysate into 250 μL aliquots (Input DNA) and store at −80 °C until ready to proceed with immunoprecipitation (*see* **Note 7**).

3.4 Immunoprecipitation

1. In two separate 1.5 mL microcentrifuge tubes combine 25 μL Protein G Sepharose (bed volume) and 100 μL of PBS.
2. Add 5 μL of Antigen-specific Serum to the first tube, and 5 μL of Control Serum to the second tube.
3. Incubate Antibody-beads with rotation at 4 °C for 1 h (*see* **Note 8**).
4. Add 500 μL of Blocking Buffer to each tube and incubate with rotation at 4 °C for 1 h.
5. Pellet Antibody-beads with centrifugation at 10,000 × *g* for 2 min.
6. Wash twice with 500 μL PBS by resuspending Protein G Sepharose and pelleting with centrifugation at 10,000 × *g* for 2 min.
7. Add 250 μL of crosslinked infected cell lysate (Input DNA) to each tube and incubate with rotation at 4 °C overnight.
8. The following day, place Wash Buffer on ice to cool.
9. Pellet Antibody-beads with centrifugation at 10,000 × *g* for 2 min.
10. Aspirate and discard supernatant.
11. Add 500 μL ice-cold Wash Buffer to each tube and incubate with rotation at room temperature 3 min.
12. Pellet Antibody-beads with centrifugation at 10,000 × *g* for 2 min.
13. Repeat **steps 10–12** three additional times for a total of four washes.
14. Prepare two new 1.5 mL microcentrifuge tubes by adding 80 μL of Neutralization Buffer to each tube.
15. Add 100 μL of Elution Buffer to Antibody-beads and incubate at room temperature for 1 min.
16. Pellet Antibody-beads with centrifugation at 10,000 × *g* for 2 min.
17. Remove supernatant and add directly to a new tube containing Neutralization Buffer (*see* **Note 9**).

18. Repeat **steps 15–17** three additional times for a total of four elutions.
19. Discard tubes containing Antibody-beads.
20. Incubate tubes containing neutralized eluate at 95 °C for 1 h to reverse crosslinks (*see* **Note 10**).
21. Allow eluate to cool to room temperature.
22. Use Nucleospin Gel and PCR Clean-up Kit according to the manufacturer's instructions to purify DNA from eluate, with a final volume of 100 μL.
23. Store purified DNA at −20 °C.
24. To determine the amount of input DNA used in each Intra-ChIP reaction, incubate a separate 250 μL aliquot of cross-linked lysate (Input DNA) at 95 °C for 1 h to reverse crosslinks.
25. Perform **steps 21–23** to obtain purified Input DNA in a final volume of 100 μL.

3.5 Quantitative Real-Time PCR

1. Prepare PCR master mixes by combining primers for Target DNA or Non-target DNA (final 250 nM concentration) with SYBR Green master mix according to the manufacturer's instructions (*see* **Note 11**).
2. Add 20 μL of master mix to optically clear PCR tubes.
3. For each primer pair, set up triplicate reactions by adding 5 μL of the following templates:
   (a) Control Serum Immunoprecipitated DNA.
   (b) Antigen-specific Serum Immunoprecipitated DNA.
   (c) Purified Input DNA.
   (d) H_2O (negative control).
4. Use a Bio-Rad iCycler to perform Real-Time PCR, with 40 cycles of 95 °C for 30 s and 60 °C for 30 s, with detection occurring during the 60 °C step.

3.6 Data Analysis

1. To facilitate direct comparison of the Ct values for Target DNA, Non-target DNA, and Input DNA, 5% of each sample is used as a template in each Real-Time PCR reaction (5 μL out of 100 μL).
2. Calculate the amount of Target DNA or Non-target DNA isolated with either Control Serum or Antigen-specific Serum relative to Purified Input DNA (Input) using the formula below (*see* **Note 12**):

$$\%\text{Input recovered} = 100\% \times 2^{\left(\text{Ct}_{\text{Input}} - \text{Ct}_{\text{ChIP}}\right)}.$$

3. Calculate Fold Enrichment of Target DNA or Non-target DNA using the formula below (*see* **Note 13**):

$$\text{Fold Enrichment} = \frac{\text{Antigen-specific Serum \% Input Recovered}}{\text{Control Serum \% Input Recovered}}.$$

4 Notes

1. The use of a biosafety cabinet is required to maintain sterility when performing tissue culture.
2. *Chlamydia trachomatis* is a human pathogen and requires a Biosafety Level 2 facility and precautions.
3. Crosslinking solution must be made up fresh just prior to adding to infected cells. A new 37% formaldehyde solution should also be prepared approximately every 1–2 months due to instability in aqueous solution.
4. After crosslinking, it is critical to quench and remove excess formaldehyde. If formaldehyde is carried over to the immunoprecipitation step, it will result in nonspecific crosslinking of protein–DNA complexes to the Antibody-beads.
5. Sonication conditions that result in DNA fragmentation between 300 and 1200 bp must be determined empirically, and are dependent on a number of factors including; sonifier, cell density, and buffer. Care must be taken to ensure the sample does not heat up during sonication, because this can result in reversal of crosslinking. Fragmentation to this size range allows for resolution of binding events to individual loci that are at least 1200 bp apart on the bacterial genome. Because both host DNA and bacterial DNA are present in the same sample, the desired size range is actually a composite of all total DNA. In our experience, achieving this size range with total DNA has proven sufficient for detection of bacterial binding events.
6. Pre-clearing lysate with Protein G Sepharose aids in removing nonspecific interactions that may occur during the immunoprecipitation step.
7. Lysates are very stable when stored at −80 °C, and we have successfully performed ChIP after over a year of storage.
8. We pre-bind antibodies to Protein G Sepharose to ensure that Protein G Sepharose is not a limiting factor. If excess antibodies are added directly to the lysate, protein–DNA interactions may be missed if there is insufficient Protein G Sepharose to bind all the antibodies.

9. The low pH of the Elution Buffer is responsible for dissociating antibodies from Protein G Sepharose, and for dissociating protein–DNA complexes from antibodies. However, this low pH can be damaging to DNA, so it is important to rapidly neutralize the eluate.
10. It is critical to reverse the protein–DNA crosslinks prior to analysis; otherwise the primers for Real-Time PCR will be unable to bind the template. It is also possible to reverse crosslinks by incubating at 65 °C overnight instead of using 95 °C. Crosslinking reversal conditions should be determined empirically by running DNA on an agarose gel with ethidium bromide staining. Successful reversal of crosslinks will result in a smear of DNA within the desired size range (300–1200 bp). Insufficient reversal of crosslinks will result in a DNA smear that extends all the way up to the well.
11. Primer design is an important part of the Real-Time PCR analysis. Primers should be designed to have a melting temperature of ~60 °C and should amplify ≤100 bp of Target or Non-target DNA. It is also important to verify that the primers only amplify a single product from Purified Input DNA, as assessed with gel electrophoresis and ethidium bromide staining. Nonspecific interactions between primers and host-cell DNA can also result in reduced amplification efficiency. Therefore, it is also important to determine amplification efficiency for each primer pair using Purified Input DNA, which contains a mix of host and bacterial DNA.
12. For the formula given, $2^{(Ct\_Input\ -\ Ct\_ChIP)}$ assumes an amplification efficiency of 100%, leading to a twofold difference in copies between each Ct value. However, primer amplification efficiency can vary significantly. Amplification efficiency can be determined by generating standard curves with DNA templates at known copy numbers, or by using linear regression analysis directly on sample data [6].
13. Fold Enrichment is used to determine the amount of DNA isolated with a specific antibody relative to a nonspecific antibody. This calculation indicates the specificity of the Intra-ChIP procedure. When used in combination with a Non-Target DNA control, Fold Enrichment accounts for antibody specificity and Target DNA specificity.

References

1. Collas P (2010) The current state of chromatin immunoprecipitation. Mol Biotechnol 45:87–100
2. Hanson BR, Tan M (2015) Transcriptional regulation of the Chlamydia heat shock stress response in an intracellular infection. Mol Microbiol 97:1158–1167
3. Hanson BR, Tan M (2016) Intra-ChIP: studying gene regulation in an intracellular pathogen. Curr Genet 62:547–551

4. AbdelRahman YM, Belland RJ (2005) The chlamydial developmental cycle. FEMS Microbiol Rev 29:949–959
5. Belland RJ, Zhong G, Crane DD, Hogan D, Sturdevant D, Sharma J, Beatty WL, Caldwell HD (2003) Genomic transcriptional profiling of the developmental cycle of Chlamydia trachomatis. Proc Natl Acad Sci U S A 100:8478–8483
6. Peirson SN, Butler JN, Foster RG (2003) Experimental validation of novel and conventional approaches to quantitative real-time PCR data analysis. Nucleic Acids Res 31:e73

Chapter 14

Native Chromatin Immunoprecipitation-Sequencing (ChIP-Seq) from Low Cell Numbers

Teodora Ribarska and Gregor D. Gilfillan

Abstract

ChIP-seq is the current method of choice for genome-wide protein location analysis. Here, we present a native (non-cross-linked) ChIP procedure suitable for histone proteins, coupled with an efficient library preparation technique for subsequent next-generation sequencing. The method enables ChIP-seq starting with 50,000 or more cells.

Key words HTS, NGS, Micro-ChIP, Next-generation sequencing, N-ChIP

1 Introduction

ChIP allows the genomic locations bound by a protein to be determined. Following formaldehyde cross-linking and mechanical shearing (X-ChIP), or enzymatic digestion of un-fixed cells (native, or N-ChIP), immuno-enrichment of the target protein also captures associated DNA. To determine binding at a single locus by quantitative PCR, ChIP protocols requiring as few as 100 cells can be performed [1]. However, to analyze genome-wide binding profiles requires higher amounts of material to allow either micro-array hybridization (ChIP-chip) or high-throughput sequencing (ChIP-seq). Whilst ChIP-chip has been performed with as few as 1000 cells [2], the advantages in sensitivity and resolution offered by ChIP-seq mean that ChIP-seq is more often favored [3, 4], despite higher input requirements. “Cutting edge” techniques currently aim to enable ChIP-seq from single cells [5, 6], but these procedures necessitate custom-built microfluidic devices, compromise sensitivity, and require copious amounts of sequencing. Thus, until methodological advances allow cheap, easily performed and sensitive assays of single cells, in most cases it will remain appropriate to rely on the averaged signals derived from ChIP assays performed on thousands of pooled cells. Native chromatin inherently

Neus Visa and Antonio Jordán-Pla (eds.), *Chromatin Immunoprecipitation: Methods and Protocols*,
Methods in Molecular Biology, vol. 1689, DOI 10.1007/978-1-4939-7380-4_14,

offers higher epitope accessibility, improving the immunoprecipitation efficiency and thereby also lowering the requirement for starting cell numbers.

The protocol presented here has been used for fresh and live-frozen cells [7]. As written, it has been optimized for live-frozen magnetically sorted $CD4^+$ or $CD8^+$ lymphocytes in the range of 2×10^5–1×10^6 cells (starting cell numbers), but can easily be adjusted to start with fresh cultured or isolated cells. Scaling all volumes down by a factor of 5 has been tested and works well down to at least 4×10^4 cells. The protocol generates chromatin that can be diluted and used for multiple immunoprecipitations (e.g., chromatin equivalent to 2×10^4 cells per IP).

Note that since this protocol is a Native-ChIP procedure [7–10], it is only recommended for histone proteins due to the use of micrococcal nuclease digestion. If another protein target is intended, an X-ChIP procedure is likely to be more appropriate.

2 Materials

2.1 Reagents

1. Micrococcal Nuclease (MNase—*see* **Note 1**): (Sigma-Aldrich, St. Louis, MO).

   Rehydrate the MNase by adding to the vial 0.5 mL of MNase storage buffer (50% Glycerol solution supplemented with 10 mM Tris–HCl, pH 7.5 and 50 mM NaCl) to a stock concentration of 0.1 U/μL, aliquot and store at −20 °C. The enzyme from this provider is particularly stable and gives reproducible digestion profiles.
2. Proteinase K.
3. Specific antibodies (ChIP-seq grade—*see* **Note 2**) and control IgG.
4. TPX 1.5 mL tubes (Diagenode s.a., Liège, Belgium; *see* **Note 3**).
5. 8-well 0.2 mL PCR strip tubes.
6. Dynabeads Protein A or G beads for immunoprecipitation (Thermo Fisher Scientific, Waltham, MA).
7. DNA purification columns.
8. Protease Inhibitor Cocktail (Sigma-Aldrich, St. Louis, MO). Use at 100× dilution (high conc.) or 1000× dilution (low conc.).
9. Reagents and primers for quantitative PCR.
10. Agencourt AMPure XP DNA purification beads (Beckman Coulter, Brea, CA).
11. 99% pure laboratory ethanol.

12. Qubit dsDNA HS Assay Kit (Thermo Fisher Scientific).
13. 2100 Bioanalyzer High Sensitivity DNA Kit (Agilent Technologies, Santa Clara, CA).
14. ThruPLEX DNA-seq Kit (Rubicon Genomics, Ann Arbor, MI; *see* **Note 4**).

2.2 Equipment

1. Cold room or large refrigerator.
2. Magnet for 1.5 mL tubes.
3. Magnet for 0.2 mL tubes.
4. Bioruptor® Plus (Diagenode s.a.; *see* **Note 3**).
5. 2100 Bioanalyzer (Agilent Technologies).
6. Microfuge for 1.5 mL tubes—room temp.
7. Microfuge for 1.5 mL tubes—chilled.
8. Mini centrifuge with rotor for 8-well strips.
9. Rotating wheel with mountings holding 8-well strips.
10. Shaking thermal block for 1.5 mL tubes.
11. Quantitative real-time PCR machine.
12. (Optional) Aspirator apparatus and waste trap.
13. Qubit Fluorimeter (Thermo Fisher Scientific).

2.3 Buffers

1. MNase Digestion buffer: 50 mM Tris–HCl, pH 8.0, 1 mM $CaCl_2$, 0.2% Triton X-100 (*see* **Note 5**).
2. 10× MNase Stop buffer: 110 mM Tris–HCl, pH 8.0, 55 mM EDTA.
3. 10× PBS: 80 g NaCl, 2 g KCl, 14.4 g Na_2HPO_4, 2.4 g KH_2PO_4. Dissolve the salts in 1 L H_2O, then autoclave. Prepare a fresh 1× stock from this prior to starting protocol below.
4. 2× RIPA IP buffer: 280 mM NaCl, 1.8% Triton X-100 (*see* **Note 5**), 0.2% SDS, 0.2% sodium deoxycholate (*see* **Note 6**), 5 mM EGTA (*see* **Note** 7). Chill on ice prior to starting procedure.
5. RIPA buffer: 10 mM Tris–HCl, pH 8.0, 1 mM EDTA, 140 mM NaCl, 1% Triton X-100, 0.1% SDS, 0.1% sodium deoxycholate. Chill on ice prior to starting procedure.
6. LiCl wash buffer: 250 mM LiCl, 10 mM Tris–HCl, pH 8.0, 1 mM EDTA, 0.5% NP-40 (now known as Igepal CA-630, *see* **Note 5**), 0.5% sodium deoxycholate. Chill on ice prior to starting procedure.
7. TE: 10 mM Tris–HCl, pH 8.0, 1 mM EDTA. Chill on ice prior to starting procedure.

3 Methods

3.1 DAY 1: Cell Harvesting, MNase Digestion, and IP Set Up

1. Thaw an aliquot of frozen cells in a water bath at 37 °C. Immediately transfer the cells to a 1.5 mL TPX tube and centrifuge for 5 min at 1500 × *g* at room temp.
2. Wash cells with PBS by resuspending cell pellet in 1 mL PBS (room temp), then centrifuge again as above and discard supernatant in hazardous waste.
3. Whilst cells are centrifuging, prepare a fresh working dilution of MNase in digestion buffer sufficient for all samples you will process. Use pipetting volumes that will be most reproducible (i.e., avoid pipetting 1 μL volumes—suggested volume used here is 5 μL). Store diluted enzyme on ice. Discard remaining diluted enzyme after use.
4. Remove and discard supernatant, then resuspend cell pellet in 95 μL digestion buffer per 1×10^5–1×10^6 cells (room temp) supplemented with protease inhibitors (high conc.). Immediately pipette cells gently up and down several times to prevent aggregation of lysed cell components.
5. For every 95 μL cell suspension, add 5 μL diluted MNase to cells, flick tube to mix, quickly spin down and then place in 37 °C block (no shaking) for exactly 5 min. Process one sample at a time to ensure exact digestion times. Note that the amount of MNase required will need to be determined empirically for each new batch of enzyme purchased (*see* **Note 1**) or for new cell types.
6. Add 10 μL 10× stop buffer, mix thoroughly but gently. Keep samples on ice/in cold room from now on.
7. Sonicate in the Bioruptor® Plus for 60 s (high power, no pulsing), or 30 s if using a Bioruptor® Pico. Volume in individual 1.5 mL tubes must not exceed 300 μL.
8. Adjust buffer to 1× RIPA conditions by adding cold 2× RIPA IP buffer (110 μL per 1×10^6 cells) supplemented with protease inhibitors (high conc.). Mix by flicking tube.
9. Spin tubes for 15 min in a microfuge, top speed, 4 °C.
10. Collect supernatants to fresh pre-cooled tubes (1.5 mL regular polypropylene) and store on ice. Pool those from the same sample (where applicable) at this point (*see* **Note 8**).
11. Remove 10% volume of chromatin and store for later purification. This sample represents the "input" chromatin, and will be used in qPCR and to control digestion on an Agilent Bioanalyzer. When convenient, digest this aliquot by adding TE to 95 μL and 5 μL proteinase K (10 mg/mL stock). Incubate with shaking (1200 rpm or greater) in a 55 °C block for 1 h. Purify

over a spin column and elute in 10 μL 10 mM Tris–HCl, pH 8. Load 1 μL on Bioanalyzer DNA HS chip as per manufacturer's instructions. Digestion should reveal a majority of mono-nucleosomes (ca. 140 bp) with di- and tri-nucleosome peaks also clearly visible (Fig. 1). Store remainder for use as "input" samples in qPCR.

12. Prepare sufficient protein A & G Dynabeads. Per IP, a total of 25 μL beads (1:1 mix of protein A and protein G—this can be mixed in advance) are required: 10 μL for preclearing, 10 μL for IP and ca. 5 μL excess. Pipette the desired amount into a 1.5 mL tube, and using the magnet, wash them twice with 1 mL RIPA buffer with protease inhibitors (low conc.). Finally, resuspend them to their original volume in RIPA buffer with protease inhibitors (low conc.).
13. Dilute chromatin to 100 μL/IP with RIPA containing protease inhibitors (low conc.). IPs will later be performed in 0.25 mL thin-walled PCR tubes, which give best mixing upon rotation with 100–150 μL volumes.
14. Pre-clear chromatin with beads: For every IP you will perform, add 10 μL washed Dynabeads to the chromatin, e.g., if you will perform 4 separate IPs, add 40 μL beads. Incubate with rotation at 4 °C for 1 h.
15. Place the tubes on a magnetic rack, wait for suspension to clear, then transfer 100 μL chromatin per IP to a well of an 8-well thin-walled tube strip. Discard the beads. Add antibodies (amounts must be determined empirically), seal with 8-well strip caps, and incubate overnight with rotation at 4 °C. Be sure to perform appropriate negative controls such as pre-immune or control IgG IPs in addition to specific immunoprecipitations.

3.2 DAY 2: IP and Verification by qPCR

1. Spin down tubes in a mini-centrifuge to collect any liquid from lids. Add 10 μL beads per IP and incubate a further 2–3 h with rotation at 4 °C.
2. Whilst beads are binding to antibody, prepare sufficient RIPA and LiCl wash buffers for all IPs, supplemented with protease inhibitors (low conc.).
3. Collect beads on magnet, and discard supernatant. Wash beads five times with 150 μL RIPA buffer and once with 150 μL LiCl wash buffer. All washes should be performed with rotation at 4 °C for 5 min in cold room, using magnet to collect beads between washes. Beads stuck in tube lids should be collected by briefly spinning tube strips in mini-centrifuge prior to placing

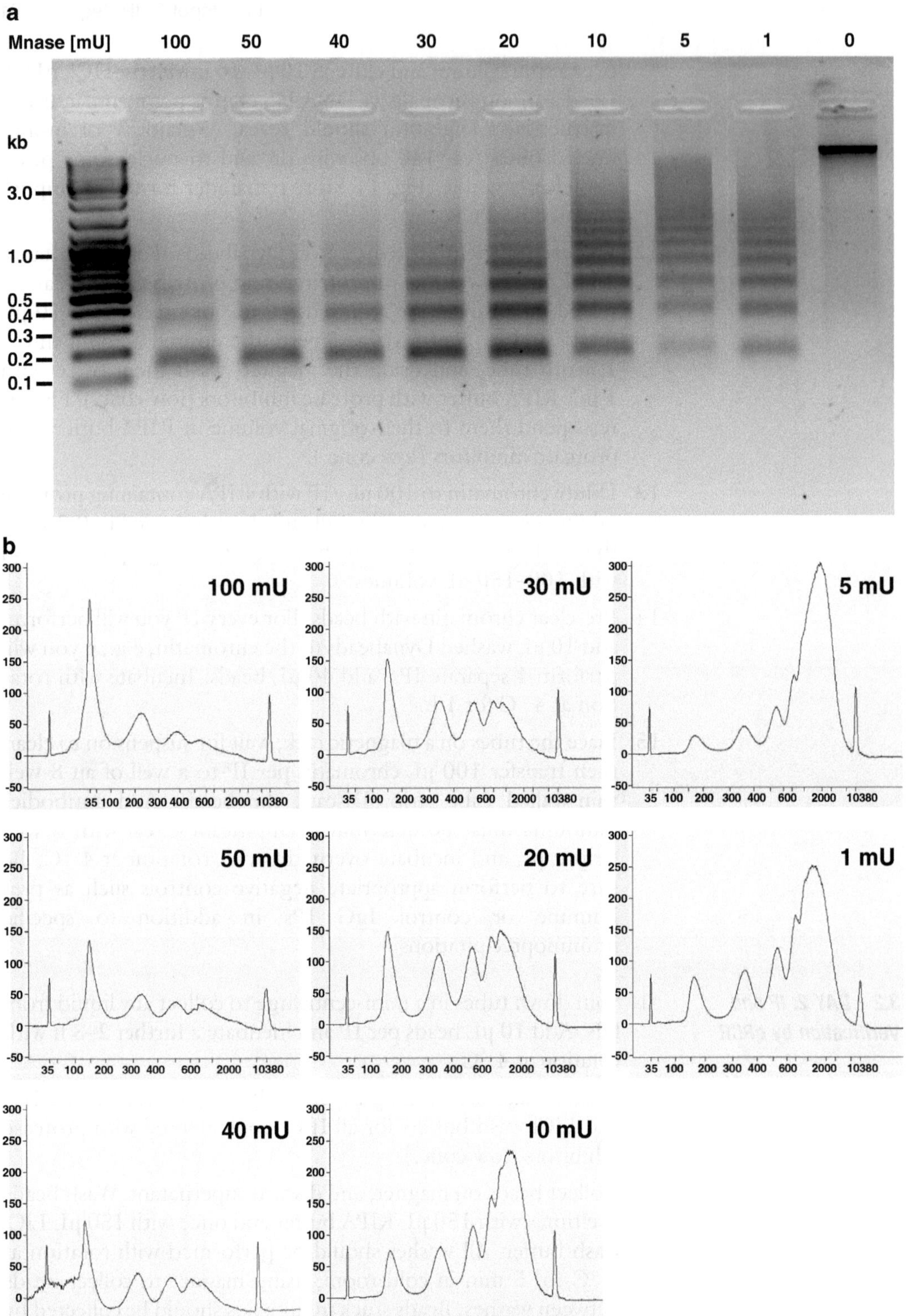

Fig. 1 DNA profiles of MNase-digested cells. One million HelaS3 cells were digested with 0, 1, 5, 10, 20, 30, 40, 50, or 100 mU MNase (Sigma). Native chromatin was isolated, treated with Proteinase K and the DNA was isolated with a PCR product isolation kit. (**a**) Gel electrophoresis using 0.5-1 μg digested DNA loaded on 1.5%

strips on magnet. Use of an aspirator to remove buffers will greatly speed this step.

4. Rinse beads with a final wash of 150 μL cold TE (no protease inhibitors). Note that beads do not adhere so well to tube walls in TE buffer, so use a pipette rather than aspirator to remove TE.
5. To digest proteins, prepare a sufficient amount of Proteinase K mix: 95 μL TE and 5 μL Proteinase K (10 mg/mL stock) per IP. Resuspend each bead pellet in 100 μL Proteinase K mix, transfer to a fresh 1.5 mL tube, and incubate with shaking (1200 rpm or greater) in a 55 °C block for 1 h.
6. Purify immunoprecipitated DNA using a clean-up column by adding column binding buffer directly to beads/TE mix, and vortex. Place tube on magnet and wait until beads aggregate, then transfer supernatant to clean-up column, and proceed as per manufacturer's instructions. Elute in 50 μL 5 mM Tris for maximum recovery. Wait 1 min after adding elution buffer before spinning.
7. Verify IPs by qPCR, using 2 μL per 10 μL PCR reaction. Be sure to leave at least 30 μL eluate for ChIP-seq library generation, so dilute an aliquot for qPCR verification analysis if necessary. Test at least one positive and one negative control region appropriate to each antibody before proceeding to prepare Illumina sequencing libraries.

3.3 DAY 3: Sequencing Library Preparation

1. Determine concentration of 1 μL of IP eluate using the Qubit HS DNA reagents and a Qubit fluorometer, following manufacturer's instructions (*see* **Note 9**). Determine also the concentration of input samples (which may need dilution to be within the range of the Qubit HS reagents) if these are also to be sequenced.
2. Prepare sequencing library using a ThruPLEX DNA-seq kit (Rubicon Genomics) or equivalent low-input reagents (*see* **Notes 10** and **11**) following the manufacturer's instructions.

Fig. 1 (continued) agarose gel. (**b**) Bioanalyzer profiles of the same digested DNA. 10 ng digested DNA was loaded per well of a Bioanalyzer DNA High Sensitivity chip according to the manufacturer's instructions and run on a Bioanalyzer 2100 instrument. The profiles generated with the five highest amounts of MNase (20–100 mU) shown are suitable for ChIP. Depending upon experimental aims, the most appropriate digestion conditions should be chosen—less digestion (e.g. 20 mU in this case) will pull down more DNA, but give a lower resolution. Overdigestion, producing only mononucleosomes, is not recommended as immunoprecipitation efficiency will be adversely affected

While library preparation is possible with amounts as low as 50 pg DNA, we recommend to use 1-10 ng DNA if possible. It is important to use the same amount from "input" DNA samples as for IP samples if this will also be sequenced (recommended), so that both input and ChIP samples undergo the same amount of amplification. Pay attention to your choice of indexes, to be sure that compatible indexes can be run together in a single lane/run if necessary (*see* **Note 12**).

3. Measure the concentration of your library by Qubit HS and check the fragment size on a Bioanalyzer HS DNA chip. The library should give a range of fragments similar to the digested input chromatin (plus the length of the linker adaptors and primers), but will be biased toward amplification of lower molecular weight products (mononucleosomes). It is important that the library does not contain a significant peak of adapter dimer at approximately 120 bp.

3.4 Sequencing

Most sequencing commercial or core-facilities will require that you supply Bioanalyzer traces and concentration information for your samples, and will perform further QC to determine optimum loading. For ChIP samples, we recommend the use of the Illumina NextSeq system with 75 cycle reagents, which can deliver approximately 400 million reads per run, sufficient to allow multiplexing of several samples per run. You can choose in this case to run 75 bp single-end reads, or 40 bp paired end reads to generate a read from each end of your ChIP DNA. For MNase-digested chromatin, using paired-end reads may help increase the number of uniquely mapping fragments, as duplicate reads are more common in MNase-digested chromatin than randomly fragmented (sonicated) DNA.

4 Notes

1. Each batch of MNase purchased may require optimization to adjust for differences in activity, so it is recommended to purchase sufficient enzyme for multiple experiments and freeze in appropriate aliquots.
2. Many commercial vendors now offer ChIP-seq-validated antibodies. Nonetheless, quality and the criteria required to meet this grade vary, and for polyclonal antibodies, batch-to-batch variation is to be expected. We recommend verifying the specificity of histone antibodies by binding to commercially available peptide arrays. Once a suitable antibody has been identified, it is advisable to purchase sufficient antibody from the same lot/batch number to complete all your planned experiments.

3. The experimental conditions detailed here are optimized for a Bioruptor® Plus sonicator with TPX plasticware, or the Bioruptor® Pico sonicator (using half the sonication time) with the recommended plasticware. Other sonicators can be used, but will require optimization of sonication conditions. The aim of the sonication in this case is not to shear the chromatin (as is the case with X-chip), but to increase the yield of soluble chromatin fragments by a milder disruption. We have found that the use of plasticware appropriate for your sonicator is essential to obtain the most reproducible results.
4. Choose the most appropriate kit for the number of reactions you will sequence, paying attention to indexing requirements (see kit guidelines for more details).
5. Triton X-100 should be added from a 10% (w/v) stock in water. For greatest accuracy weigh the liquid Triton (density 1.07 g/mL). Prepare 10% Igepal CA-630 in a similar manner.
6. Sodium deoxycholate 10% stock: filter-sterilize after dissolving in water.
7. EGTA dissolves in water at pH 8. Use 10 N NaOH to adjust the pH and prepare 0.5 M stock EGTA.
8. The native chromatin cannot be frozen or snap-frozen, it must always be freshly prepared and used for IP. If cells need to be collected at different time points, we recommend snap-freezing (in liquid nitrogen) the washed cells as liquid-free pellets at 1×10^6 cells/tube and storing them at −80 °C until they can be processed into chromatin at a later time point. Frozen pellets can be thawed on ice when desired and resuspended directly in MNase digestion buffer (Subheading 3.1, **step 4**).
9. Even if you are unable to detect immunoprecipitated DNA using the Qubit HS reagents (i.e., your samples are below detection limits), it may still be possible to generate a sequencing library, in which case the maximum recommended number of PCR cycles should be selected when using the ThruPLEX library preparation reagents.
10. To ensure that your IP conditions capture sufficient DNA to reproducibly prepare sequencing libraries and gain experience, we recommend you to prepare your own libraries. Once you are confident of the quality, outsourcing library preparation to a core facility may save resources, but you will generally receive data in return much faster if you are able to submit completed sequencing libraries.
11. The ThruPLEX reagents recommended here were selected for use in our facility following a comparative study of seven low-

input library preparation methods [11], and offer excellent performance combined with an efficient and easily followed protocol.

12. Most library preparation kits contain a section explaining index compatibility, but we reiterate the point here that in the case of Illumina sequencing, index sequencing reads should contain indexes in combination such that each index sequencing cycle contains at least one base from each of the following two pairs: (G/T) and (A/C).

Acknowledgment

G.D.G. is funded by the Norwegian South-Eastern Regional Health Authority. T.R. was funded by the EU-FP7 project EpiTrain (REA grant agreement n° 316758). We would also like to thank Diagenode for their support.

References

1. Dahl JA, Klungland A (2015) Micro chromatin immunoprecipitation (muChIP) from early mammalian embryos. Methods Mol Biol 1222:227–245
2. Dahl JA, Reiner AH, Collas P (2009) Fast genomic muChIP-chip from 1,000 cells. Genome Biol 10(2):R13
3. Furey TS (2012) ChIP-seq and beyond: new and improved methodologies to detect and characterize protein-DNA interactions. Nat Rev Genet 13(12):840–852
4. Ho JW, Bishop E, Karchenko PV et al (2011) ChIP-chip versus ChIP-seq: lessons for experimental design and data analysis. BMC Genomics 12:134
5. Cao Z, Chen C, He B et al (2015) A microfluidic device for epigenomic profiling using 100 cells. Nat Methods 12(10):959–962
6. Rotem A, Ram O, Shoresh N et al (2015) Single-cell ChIP-seq reveals cell subpopulations defined by chromatin state. Nat Biotechnol 33(11):1165–1172
7. Gilfillan GD, Hughes T, Sheng Y et al (2012) Limitations and possibilities of low cell number ChIP-seq. BMC Genomics 13:645
8. O'Neill LP, Turner BM (2003) Immunoprecipitation of native chromatin: NChIP. Methods 31(1):76–82
9. Thorne AW, Myers FA, Hebbes TR (2004) Native chromatin immunoprecipitation. Methods Mol Biol 287:21–44
10. Umlauf D, Goto Y, Feil R (2004) Site-specific analysis of histone methylation and acetylation. Methods Mol Biol 287:99–120
11. Sundaram AY, Hughes T, Biondi S et al (2016) A comparative study of ChIP-seq sequencing library preparation methods. BMC Genomics 17(1):816

Chapter 15

MOBE-ChIP: Probing Cell Type-Specific Binding Through Large-Scale Chromatin Immunoprecipitation

Shenqi Wang and On Sun Lau

Abstract

In multicellular organisms, the initiation and maintenance of specific cell types often require the activity of cell type-specific transcriptional regulators. Understanding their roles in gene regulation is crucial but probing their DNA targets in vivo, especially in a genome-wide manner, remains a technical challenge with their limited expression. To improve the sensitivity of chromatin immunoprecipitation (ChIP) for detecting the cell type-specific signals, we have developed the Maximized Objects for Better Enrichment (MOBE)-ChIP, where ChIP is performed at a substantially larger experimental scale and under low background conditions. Here, we describe the procedure in the study of transcription factors in the model plant *Arabidopsis*. However, with some modifications, the technique should also be implemented in other systems. Besides cell type-specific studies, MOBE-ChIP can also be used as a general strategy to improve ChIP signals.

Key words Chromatin immunoprecipitation (ChIP), MOBE-ChIP, Cell type-specific transcriptional regulator, Protein–DNA interactions, ChIP-Seq

1 Introduction

Chromatin immunoprecipitation (ChIP) is a powerful technique for studying protein–DNA interactions in vivo, and has been widely adopted in various model organisms such as yeast, flies, and *Arabidopsis* [1–4]. However, when probing for the binding of cell type-specific transcriptional regulators, which generally express at a low level and in limited cells, the sensitivity of standard ChIP assays is often suboptimal for the detection. We recently developed an optimized ChIP technique called the Maximized Objects for Better Enrichment (MOBE)-ChIP, which significantly improves ChIP sensitivity and allows cell type-specific studies [5–7]. In MOBE-ChIP, the two key requirements are the generation of a ChIP system with minimal background noise and performing the experiment at a substantially larger scale. We attribute the improvement of our method to the disproportional increase of target signals over

Neus Visa and Antonio Jordán-Pla (eds.), *Chromatin Immunoprecipitation: Methods and Protocols*,
Methods in Molecular Biology, vol. 1689, DOI 10.1007/978-1-4939-7380-4_15,

background noise at high experimental scale when the background of a ChIP system is low (*see* [7] for a discussion on the principle). With MOBE-ChIP, we have observed an improvement on target enrichment at over 100-fold compared to standard ChIP method for a cell type-specific transcriptional regulator in plants, thus allowing the genome-wide profiling of its binding targets at high resolution [5]. MOBE-ChIP can also be used as a general strategy to improve ChIP signals or to confirm weak binding and non-binding events where enhanced target signal is especially favorable.

In this chapter, we present our MOBE-ChIP protocol which uses 16 times more starting materials than conventional ChIP assays in plants. Please refer to [7] for a detailed discussion on generating a low background system and determining the experimental scale. Briefly, low background noise is achieved by epitope-tagging and the use of highly specific antibodies, and before scaling up for MOBE-ChIP, we recommend performing ChIPs with standard protocols to test the signal strength. Similar to standard ChIP protocols [3, 4], MOBE-ChIP involves four major steps: sample cross-linking, chromatin isolation and sonication, immunoprecipitation, and reverse cross-linking. With the substantially larger starting materials, samples are divided into smaller aliquots during cross-linking and chromatin isolation steps to maintain their efficiency. Individual aliquots are only pooled at the immunoprecipitation step. This protocol was developed for *Arabidopsis* seedlings, but the principle should be applicable to other model systems. The cross-linking and chromatin isolation steps are likely specific to the biological system of interests but the remaining steps could be performed as described [7].

2 Materials

Prepare all buffers with nuclease-free ultrapure water and analytical grade reagents. Diligently follow the safety instructions when handling hazardous materials. Comply with all waste disposal regulations when disposing waste materials. For simplicity, the amount of materials needed for one sample is shown, but a negative control should normally be included (*see* **Note 1**).

2.1 Cross-Linking

1. Cross-linking buffer: 0.4 M Sucrose, 10 mM Tris–HCl, pH 8.0, 10 mM $MgCl_2$, 1% (w/v) formaldehyde. Freshly add formaldehyde (stock: 37% w/v) before each experiment. Prepare at least 350 ml cross-linking buffer and store at 4 °C (*see* **Note 2**).
2. Glycine (2 M): Weigh 3.753 g glycine and transfer to a 50-ml falcon tube. Add water to a volume of 25 ml. Store at 4 °C.
3. Vacuum chamber.

2.2 Chromatin Isolation

1. Extraction buffer 1: 10 mM Tris–HCl, pH 8.0, 0.4 M Sucrose, 10 mM $MgCl_2$, 5 mM β-mercaptoethanol. At the start of each experiment, add β-mercaptoethanol and dissolve one cOmplete protease inhibitor tablet (Roche Diagnostics, Germany) per 50 ml of buffer (or replace the tablet with a cocktail of protease inhibitors). Prepare at least 250 ml of extraction buffer 1. For storage, filter-sterilize the buffer (without β-mercaptoethanol and protease inhibitors) and store at 4 °C.
2. Extraction buffer 2: 0.25 M Sucrose, 10 mM Tris-HCl, pH 8.0, 10 mM $MgCl_2$, 1% (v/v) Triton X-100, 5 mM β-mercaptoethanol. At the start of each experiment, add β-mercaptoethanol and dissolve one cOmplete mini protease inhibitor tablet (Roche Diagnostics, Germany) per 10 ml of buffer. Prepare at least 10 ml of extraction buffer 2. For storage, filter-sterilize the buffer (without β-mercaptoethanol and protease inhibitors) and store at 4 °C.
3. Extraction buffer 3: 1.7 M Sucrose, 10 mM Tris–HCl, pH 8.0, 2 mM $MgCl_2$, 0.15% (v/v) Triton X-100, 5 mM β-mercaptoethanol. At the start of each experiment, add β-mercaptoethanol and dissolve one cOmplete mini protease inhibitor tablet (Roche Diagnostics, Germany) per 10 ml of buffer. Prepare at least 6 ml of extraction buffer 3. For storage, filter-sterilize the buffer (without β-mercaptoethanol and protease inhibitors) and store at 4 °C.
4. Nuclei lysis buffer: 50 mM Tris–HCl, pH 8.0, 10 mM EDTA and 1% (w/v) SDS. At the start of each experiment, dissolve one cOmplete mini protease inhibitor tablet (Roche Diagnostics, Germany) per 10 ml of buffer and pre-chill the buffer to 4 °C. Prepare at least 2 ml of nuclei lysis buffer. For storage, filter-sterilize the buffer (without the protease inhibitor) and store at room temperature.
5. Mortar and Pestle.
6. Tube rotator.
7. Refrigerated centrifuges for 1.5- and 50-ml tubes.
8. Miracloth.

2.3 Immunoprecipitation

1. ChIP dilution buffer: 16.7 mM Tris–HCl, pH 8.0, 167 mM NaCl, 1.2 mM EDTA and 1.1% (v/v) Triton X-100. At the start of each experiment, dissolve one cOmplete mini protease inhibitor tablet (Roche Diagnostics, Germany) per 10 ml of buffer. Prepare at least 20 ml of ChIP dilution buffer and prechilled on ice.
2. Low salt wash buffer: 20 mM Tris–HCl, pH 8.0, 2 mM EDTA, 150 mM NaCl, 0.1% (w/v) SDS, 1% (v/v) Triton X-100.

Prepare at least 8 ml of low salt wash buffer and pre-chilled on ice. For storage, filter-sterilize the buffer and store at 4 °C.

3. High salt wash buffer: 20 mM Tris–HCl, pH 8.0, 2 mM EDTA, 500 mM NaCl, 0.1% (w/v) SDS, 1% (v/v) Triton X-100. Prepare at least 8 ml of high salt wash buffer and pre-chilled on ice. For storage, filter-sterilize the buffer and store at 4 °C.
4. LiCl wash buffer: 10 mM Tris–HCl, pH 8.0, 1 mM EDTA, 250 mM LiCl, 1% (v/v) IGEPAL CA-630 (or NP-40), 0.5% (w/v) sodium deoxycholate (*see* **Note 3**). Prepare at least 8 ml of LiCl wash buffer and pre-chilled on ice. For storage, filter-sterilize the buffer and store at 4 °C.
5. TE buffer: 10 mM Tris–HCl, pH 8.0, 1 mM EDTA. Prepare at least 8 ml of TE buffer and pre-chilled on ice. For storage, filter-sterilize the buffer and store at 4 °C.
6. ChIP elution buffer: 0.1 M $NaHCO_3$, 1% (w/v) SDS. Dissolve 42 mg of $NaHCO_3$ in 4.75 ml of water. Add 250 μl of 20% (w/v) SDS to achieve the final concentration. Freshly prepare the buffer for each experiment.
7. Magnetic beads.
8. Magnetic stand.
9. Tube rotator.
10. Heat block or thermomixer.

2.4 Reverse Cross-Linking

1. NaCl (5 M): Weigh 7.3 g NaCl and transfer to a 50-ml falcon tube. Add water to a volume of 25 ml.
2. DNA purification kit.
3. Protease buffer: 385 mM Tris–HCl, pH 6.5, 96 mM EDTA, 0.77 mg/μl of Proteinase K. Prepare at least 210 μl of protease buffer and keep at room temperature. Prepare freshly before the experiment.
4. Heat block or thermomixer.

3 Methods

The following MOBE-ChIP procedure is illustrated with a single sample. However, in most cases, at least an additional negative control would be needed and they should be processed in parallel (*see* **Note 1**) [7].

3.1 Tissue Harvesting and Cross-Linking of Plant Materials

1. Harvest plant materials in eight 50-ml centrifugation tubes with roughly 3 g (fresh weight) per tube (i.e., 24 g in total) (*see* **Note 4**). Add 37 ml cross-linking buffer to each tube and submerge the tissues (*see* **Note 5**).

2. Infiltrate the samples in a vacuum chamber for 5 min (*see* **Note 6**). At the end of the incubation, slowly release the vacuum. Invert the tubes to re-submerge the tissue and repeat the vacuum infiltration for an additional 5 min.
3. Add 2.5 ml of 2 M glycine to each tube to stop the cross-linking reaction (final concentration of glycine: 0.125 M). Mix by inverting and vacuum infiltrate for an extra 5 min.
4. Collect all the tissues in a strainer. Rinse the tissues with 500 ml to 1 l of distilled water.
5. Blot the samples with paper towels and remove as much water as possible. Flash freeze the tissues with liquid nitrogen. Afterward, the cross-linked tissues can be stored at −80 °C until needed (*see* **Note 7**).

3.2 Isolation and Sonication of Chromatin

1. Grind the frozen tissue into fine powder in liquid nitrogen with pre-cooled mortar and pestle (*see* **Note 8**). Divide the powder into 6 aliquots in 50-ml centrifugation tubes (*see* **Note 9**). Resuspend each with 40 ml of pre-chilled extraction buffer 1 (4 °C). Mix the samples for 10 min on a rotator with a speed around 15 rpm at 4 °C.
2. Pre-chill six new 50-ml centrifugation tubes on ice and filter the mixture from **step 1** with two layers of Miracloth into the new centrifugation tubes (*see* **Note 10**).
3. Centrifuge the filtered solution at 3000 × *g* at 4 °C for 20 min.
4. Carefully remove the supernatant with a 25-ml pipette without disturbing the pellet. Add 1.3 ml of pre-chilled extraction buffer 2 (4 °C) to each tube and resuspend the pellet by pipetting up and down (*see* **Note 11**).
5. Transfer the resuspension to six new 1.5-ml microcentrifuge tubes. Centrifuge at 12,000 × *g* at 4 °C for 10 min.
6. Carefully remove the supernatant with a pipette without disturbing the pellet. Add 400 μl of pre-chilled extraction buffer 3 (4 °C) to each tube and resuspend the pellet by gently pipetting up and down (*see* **Note 12**).
7. Prepare six new 1.5-ml microcentrifuge tubes containing 400 μl of pre-chilled extraction buffer 3 (4 °C). Overlay the buffer with the resuspensions from **step 6**. Centrifuge the six tubes at 16,000 × *g* at 4 °C for 1 h.
8. Discard the supernatant and add 300 μl of pre-chilled nuclei lysis buffer (4 °C). Resuspend the pellet by gently pipetting up and down (*see* **Note 13**). Incubate on ice for 10 min.
9. Sonicate the resuspension with a sonicator (*see* **Note 14**). If using a Bioruptor® sonicator, the following condition may be used as a reference point: three 7.5-min rounds with a 30 s ON

and a 30 s OFF cycle (total ON cycles: 27) at high output intensity (*see* **Notes 15** and **16**).

10. Centrifuge the sonicated samples at 12,000 × *g* at 4 °C for 10 min.
11. Transfer the supernatants to a pre-chilled 2-ml microcentrifuge tube (total volume: around 1.8 ml) (*see* **Note 17**). Aliquot 10 μl of the pooled sample as an Input control in following steps and store at −20 °C. The procedure can be paused here and the rest of the sonicated chromatin can be stored at −80 °C.

3.3 Immunoprecipitation of Chromatin

1. Dilute the combined chromatin solution ten times with ChIP dilution buffer (4 °C) to a final volume of around 17 ml in a 50-ml centrifugation tube.
2. Add 90 μg of antibody to the diluted chromatin and incubate the sample overnight (or >5 h) at 4 °C using a rotator (*see* **Notes 18** and **19**).
3. Add 400 μl of pre-washed magnetic beads, such as Dynabeads® Protein A (*see* **Note 20**). Rotate the samples at 4 °C for 1 h.
4. Capture the magnetic beads with a magnetic stand on ice. Remove and discard the supernatant with a pipette.
5. Resuspend the magnetic beads with 8 ml of pre-chilled low salt wash buffer (4 °C). Wash the beads at 4 °C for 5 min on a rotator. Spin the tube(s) briefly (<3800 × *g* and ~10 s) and place on a magnetic stand. Discard the supernatant with a pipette.
6. Wash the beads with pre-chilled high salt wash buffer, LiCl wash buffer and TE buffer (one wash per buffer). For each wash, incubate on a rotator at 4 °C for 5 min, spin the tube(s) briefly (<3800 × *g* and ~10 s), place them on a magnetic stand, and discard the supernatant with a pipette.
7. Add 1.6 ml of ChIP elution buffer (room temperature) to the captured magnetic beads. Vortex to mix and incubate at 65 °C for 40 min. Mix every 5 min during the incubation.
8. Spin the tube(s) briefly (<3800 × *g* and ~10 s) and place them on a magnetic stand. Save the supernatant as eluates in a new 2-ml microcentrifuge tube.

3.4 Reverse Cross-Linking and DNA Purification

1. Retrieve the Input control from **step 11** in Subheading 3.2 and let it thaw completely on ice. Add 190 μl of ChIP elution buffer (room temperature) and 8 μl of 5 M NaCl to the Input sample and mix well.
2. Add 64 μl of 5 M NaCl to the 1.6 ml eluates and mix well.
3. Incubate both the Input sample and eluates at 65 °C for at least 6 h to overnight.

4. Add 166.4 and 41.6 μl of protease buffer (room temperature) to the eluates and the Input sample, respectively. Incubate at 45 °C for 1 h.
5. Purify the de-crosslinked DNA using a column-based DNA purification kit, such as ChIP DNA Clean & Concentrator™ (Zymo, USA). The total sample volume will be larger than the column capacity. Add aliquots of the sample into the same column repeatedly until the entire sample has been processed (*see* **Note 21**).
6. Add 15 μl of elution buffer (from the DNA purification kit) to the column and incubate at room temperature for 1 min (*see* **Note 22**). Repeat the elution once with the same column. Store the purified DNA at −20 °C.
7. Quantify the DNA yield of the samples using fluorescence-based methods, such as using the Qubit® fluorometer (ThermoFisher Scientific, USA) (*see* **Note 23**). This is recommended if the samples will be used for the construction of a ChIP-Seq library.
8. Test the samples for the enrichment of specific genomic regions by quantitative PCR to determine the success of the ChIP assay (*see* **Note 24**).

4 Notes

1. It is highly recommended to include a negative control. For epitope-tagged samples, a non-tagged or wild-type control should be included. For direct pull-down of the endogenous transcriptional regulator of interests with its specific antibody, a mutant that does not express the regulator is recommended. The sample and negative control should be processed in parallel.
2. Wear personal protective equipment and handle formaldehyde with care in a fume hood, as it is flammable and poisonous. Also, avoid using formaldehyde that is opened for more than one year, as air promotes its oxidation.
3. Sigma-Aldrich has replaced NP-40 with IGEPAL CA-630, which is declared to be a chemically duplicate of NP-40. In this experiment, either can be used.
4. The amount of starting materials varies based on the type of tissues used. For example, a much higher amount is typically used when processing young seedling samples, while less would be needed for inflorescences. In addition, in determining the scale of the ChIP experiment, it is advisable to conduct a pilot ChIP assay using standard established protocols specific to the model system.

5. Samples are divided into standard-sized aliquots to maintain the cross-linking efficiency. Overloading with extra samples may lead to insufficient cross-linking.
6. The indicated infiltration time (total: 10 min) only serves as a reference point. Excessive cross-linking may lead to inefficient sonication. Hence, optimization on the cross-linking time and the concentration of formaldehyde are needed for different tissue amount and types.
7. Avoid thawing the sample if the procedure is paused.
8. Avoid thawing the samples during grinding. Add additional liquid nitrogen if needed to keep samples frozen at all times.
9. Standard-sized aliquots are used to maintain efficiency of the nuclei isolation steps.
10. At the end of the step, gently press on the miracloth to recover the remaining sample.
11. Avoid foaming when pipetting the samples.
12. With its high sucrose content, extraction buffer 3 is quite viscous. Pipette slowly and avoid vortexing the tubes to prevent foaming of the samples.
13. SDS may precipitate when nuclei lysis buffer is kept cold. Hence, right before this step, chill an aliquot of the nuclei lysis buffer on ice. Avoid vortexing the tubes.
14. For ChIP-qPCR assays, a standard probe-based sonicator should be sufficient. However, for genome-wide studies, sonicators dedicated for next-generation sequencing, such as the Bioruptor®, might be needed to generate the desired fragment size in a consistent manner.
15. Temperature of the samples should be maintained during the sonication process (~4 °C and <10 °C at all times). On a Bioruptor® sonicator without a water cooler, for example, crushed ice should be added to the water bath every 7.5 min.
16. Sonication conditions should be optimized empirically to obtain a suitable DNA profile. For general downstream analysis, DNA fragments should be less than 1 kb. For the construction of ChIP-Seq library, however, a DNA profile from 200 to 600 bp with a peak at the lower end would be ideal. Size distribution of the sonicated DNA can be evaluated by analyzing the de-crosslinked Input controls (**step 11** in Subheading 3.2) through gel electrophoresis or a BioAnalyzer. If the size of DNA exceeds 1 kb, sonication cycles can be increased. In addition, cross-linking conditions may also need to be optimized.

17. Avoid pipetting up the precipitated materials when transferring the supernatant. If insoluble materials still persist, an additional round of centrifugation may be necessary.
18. Highly specific antibody is critical in minimizing the background noise in MOBE-ChIP. It is highly recommended to evaluate the specificity of the target antibody with Western blotting and/or immunoprecipitation before the MOBE-ChIP experiment.
19. The amount of antibody for the immunoprecipitation will need to be optimized empirically. Typically, a dilution of 1:200 or a concentration of 5 μg/ml represent a good starting point. Also, based on the experimental scale, the amount of antibody, magnetic beads and buffers should be adjusted accordingly.
20. Wash the magnetic beads according to the manufacturer's protocol. Resuspend the beads in pre-chilled ChIP dilution buffer (4 °C).
21. Use only one column for each sample to reduce sample loss and ensure low sample volume for easy downstream handling. Several rounds of centrifugation are required to apply all the sample to the column.
22. The type, as well as the volume, of the elution buffer can be altered to fit with the downstream applications. The elution buffer from DNA purification kits generally help prevent DNA degradation. For constructing a ChIP-Seq library, however, it is recommended to confirm its compatibility with the library construction kit. Nuclease-free water can also be used for elution but the eluted DNA needs to be kept cold and stored at −20 °C.
23. Since highly specific antibody is used, the yield of DNA may be very low (e.g., less than 1 ng). However, we recommend quantitating the yield in a trial run to test if ChIP-Seq is feasible and if multiple MOBE-ChIPs are needed. Typically, 1 to 10 ng of DNA is required for ChIP-Seq library construction, but newer commercial kits offer input requirement at the sub-nanogram level (≥100 pg). Nevertheless, we recommend using a larger amount of DNA (>1 ng) from multiple ChIPs to achieve high complexity of the sequencing library. As an example, we pooled the eluted DNA from six MOBE-ChIPs for the ChIP-Seq experiment of a cell type-specific transcription factor [5].
24. Pilot experiments with standard ChIP protocol are recommended to identify positive and negative control regions, which would be useful to assess the success of MOBE-ChIP. As a guide, we generally observe an improvement in enrichment from three to ten-fold at the positive regions when the

starting materials are doubled using MOBE-ChIP. For the qPCR validation, 0.5 µl of DNA from the 30 µl elution would be sufficient for one qPCR reaction.

Acknowledgments

This work was supported by funds from the National University of Singapore to O.S.L. (AcRF Tier 1 R-154-000-A16-114, and start-up funds R154000695133 and R154000A06651). The authors declare no competing financial interests.

References

1. Solomon MJ, Larsen PL, Varshavsky A (1988) Mapping protein-DNA interactions in vivo with formaldehyde: evidence that histone H4 is retained on a highly transcribed gene. Cell 53:937–947
2. Orlando V (2000) Mapping chromosomal proteins in vivo by formaldehyde-crosslinked-chromatin immunoprecipitation. Trends Biochem Sci 25:99–104
3. Gendrel A-V, Lippman Z, Martienssen R, Colot V (2005) Profiling histone modification patterns in plants using genomic tiling microarrays. Nat Methods 2:213–218
4. Bowler C, Benvenuto G, Laflamme P et al (2004) Chromatin techniques for plant cells. Plant J 39:776–789
5. Lau OS, Davies KA, Chang J et al (2014) Direct roles of SPEECHLESS in the specification of stomatal self-renewing cells. Science 345:1605–1609
6. Matos JL, Lau OS, Hachez C et al (2014) Irreversible fate commitment in the Arabidopsis stomatal lineage requires a FAMA and RETINOBLASTOMA-RELATED module. Elife 3:e03271
7. Lau OS, Bergmann DC (2015) MOBE-ChIP: a large-scale chromatin immunoprecipitation assay for cell type-specific studies. Plant J 84:443–450

Chapter 16

Multiplexed ChIP-Seq Using Direct Nucleosome Barcoding: A Tool for High-Throughput Chromatin Analysis

Christophe D. Chabbert, Sophie H. Adjalley, Lars M. Steinmetz, and Vicent Pelechano

Abstract

Chromatin immunoprecipitation followed by sequencing (ChIP-Seq) or microarray hybridization (ChIP-on-chip) are standard methods for the study of transcription factor binding sites and histone chemical modifications. However, these approaches only allow profiling of a single factor or protein modification at a time.

In this chapter, we present Bar-ChIP, a higher throughput version of ChIP-Seq that relies on the direct ligation of molecular barcodes to chromatin fragments. Bar-ChIP enables the concurrent profiling of multiple DNA–protein interactions and is therefore amenable to experimental scale-up, without the need for any robotic instrumentation.

Key words ChIP-Seq, Nucleosomes, Chromatin, Multiplex, Immunoprecipitation, MNase, NGS, *Saccharomyces cerevisiae*

1 Introduction

Chromatin immunoprecipitation (ChIP) is the standard method to analyze the interactions between specific proteins or histone modifications and genomic regions. ChIP experiments coupled with microarray hybridization (ChIP-on-chip) or high-throughput sequencing (ChIP-Seq) have offered an unprecedented view of the genome-wide organization of chromatin in many organisms, and have been instrumental in our understanding of epigenetic landscapes, regulation of gene expression, and chromosome dynamics.

Classical ChIP-Seq approaches only allow profiling of a single factor or protein modification at a time. However, as our understanding of chromatin organization increases, more complex

---

Christophe D. Chabbert and Sophie H. Adjalley contributed equally to this work.

Neus Visa and Antonio Jordán-Pla (eds.), *Chromatin Immunoprecipitation: Methods and Protocols*, Methods in Molecular Biology, vol. 1689, DOI 10.1007/978-1-4939-7380-4_16, © Springer Science+Business Media LLC 2018

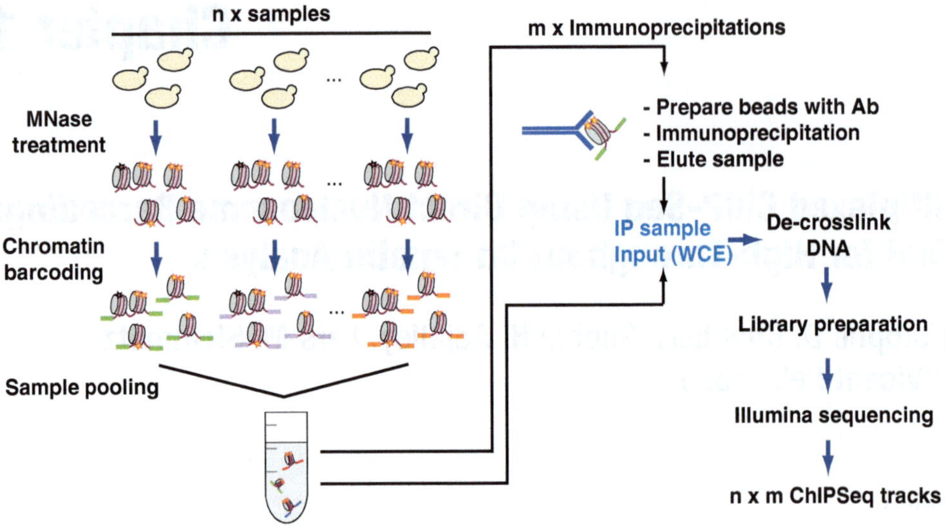

Fig. 1 Conceptual outline and experimental workflow for the Bar-ChIP approach. "n" number of samples are treated independently to produce individually barcoded chromatin fragments. Samples are pooled and "m" immunoprecipitations are conducted. The same input (WCE) is used as a control for all immunoprecipitations. After reverse-crosslinking, DNA library preparation and sequencing, "n × m" ChIP-Seq libraries can be reconstructed using bioinformatics tools

experimental designs are required. For instance, current studies often necessitate the simultaneous analysis of multiple biological samples and factors [1–9]. The continuous decrease of high-throughput sequencing costs favors the use of parallel pipelines such that the same sequencing power is distributed across multiple samples. However, ChIP remains a laborious protocol and without the use of robotic instrumentations may represent a bottleneck for the generation of genome-wide protein binding profiles.

To address these limitations, we developed a method allowing the simultaneous processing of multiple ChIP-Seq samples with limited hands on time and without the need for robotic tools [1]. Bar-ChIP (barcoded high-throughput ChIP-Seq) relies on the direct barcoding of chromatin-derived mono-nucleosome molecules (Fig. 1) prior to immunoprecipitation. This greatly reduces the handson work as it allows to simultaneously immunoprecipitate different samples that are therefore directly comparable. Experimentally, chromatin is extracted from the cells and mono-nucleosomal fractions are generated by Micrococcal Nuclease digestion (MNase). Resulting nucleosomes are barcoded with sample-specific adaptors that are compatible with Illumina sequencing. Samples are then pooled and subjected to immunoprecipitation. Barcodes introduced during the nucleosome preparation enable ChIP-Seq reads to be extracted bioinformatically and analyzed for each individual sample. Here we provide a detailed protocol of the Bar-ChIP approach [1], which has been optimized for the study of histone post-transcriptional modifications in *Saccharomyces cerevisiae* and may be adapted to other types of samples.

2 Materials

2.1 Chromatin Crosslinking

1. Formaldehyde (37%). Formaldehyde is highly toxic (*see* **Note 1**).
2. 2.5 M Glycine, diluted in water. Autoclave and keep sterile to prevent microbial contamination.
3. Tris-buffered saline (TBS) buffer: 20 mM Tris pH 7.5, 145 mM NaCl. For 1 L mix: 20 mL of 1 M Tris pH 7.5, 29.2 mL of 5 M NaCl; complete with distilled water to 1 L.

2.2 Micrococcal Nuclease Digestion

1. NP-S buffer: 0.5 mM Spermidine, 0.075% (v/v) Tergitol® NP40 (Sigma), 10 mM Tris–HCl pH 7.4, 50 mM NaCl, 5 mM $MgCl_2$, 1 mM $CaCl_2$, and 1 mM β-Mercaptoethanol. For a 200 mL stock mix 14.4 mg Spermidine, 150 μL Tergitol® NP40, 2 mL of 1 M Tris–HCl pH 7.4, 2 mL of 5 M NaCl, 1 mL of 1 M $MgCl_2$, 200 μL of 1 M $CaCl_2$, and 14 μL β-Mercaptoethanol (98%). Store at 4 °C.
2. EDTA-free protease inhibitor cocktail (Roche). Add to working volume of NP-S buffer prior to use (1 tablet per 50 mL of buffer).
3. Acid-washed glass beads (425–600 μm).
4. Micrococcal Nuclease (*see* **Note 2**).
5. 0.5 M EGTA (ethylene glycol tetraacetic acid).

2.3 Assess MNase Treatment Quality

1. TE buffer: 10 mM Tris–HCl, 1 mM EDTA.
2. RNase cocktail (Ambion) or equivalent.
3. 10% SDS, diluted in water.
4. 20 mg/mL Proteinase K.
5. 3 M NaOAC pH 5.2 for nucleic acid precipitation.
6. Ice-cold 100% and 70% (v/v) Ethanol.
7. EB buffer: 10 mM Tris–HCl pH 8.5.

2.4 Antibody Conjugation

1. Magnetic beads suitable for immunoprecipitation, such as Dynabeads Protein G or Protein A for immunoprecipitation (Invitrogen).
2. PBS/BSA: PBS buffer (137 mM NaCl, 2.7 mM KCl, 10 mM Na_2HPO_4, 1.8 mM KH_2PO_4, pH 7.4) supplemented with 5 mg/mL of BSA. Solution may be filtered and stored at 4 °C to prevent contamination.
3. Appropriate magnetic stand.

4. Tube rotator.
5. Yeast tRNA (10 mg/mL).
6. Antibodies of interest (*see* **Note 3**).

2.5 Chromatin Barcoding

Different approaches may be employed to barcode chromatin. Here, we used reagents and adapted protocols provided by Illumina and New England Biolabs (NEB).

1. *End repair step*: NEBNext End Repair Enzyme Mix and NEBNext End Repair Reaction Buffer from the NEBNext End Repair Module (NEB).
2. *A-tailing step*: Klenow fragment 3′–5′ exo- (NEB) and dATP 100 mM.
3. *Adaptors ligation step*: T4 DNA ligase buffer 10× and T4 DNA ligase (NEB).
4. *Preparation of barcoded, forked adapters.* Mix the oligos at 10 μM in the presence of 40 mM Tris–HCl pH 8.0 and 50 mM NaCl (e.g., mp1PE1 with mp1PE2, mp2PE1 with mp2PE2, etc.). Incubate sample for 5 min at 95 °C and let it cool slowly (−0.1 °C/s) to 65 °C. Incubate sample for 5 min at 65 °C and let it cool slowly (−0.1 °C/s) to 4 °C. Store the annealed linkers at −20 °C. Always thaw the linkers on ice to prevent denaturation (*see* **Note 4**).

2.6 Immunoprecipitation

1. PBS/BSA buffer: PBS buffer (137 mM NaCl, 2.7 mM KCl, 10 mM Na_2HPO_4, 1.8 mM KH_2PO_4, pH 7.4) supplemented with 5 mg/mL of BSA (Bovine Serum Albumin). Solution may be filtered and stored at 4 °C to avoid contamination.
2. Lysis buffer: 50 mM HEPES pH 7.5, 140 mM NaCl, 1 mM EDTA, 1% Triton X-100, 0.1% (w/v) Sodium Deoxycholate, 1 mM PMSF, 1 mM Benzamidin, Protease inhibitor cocktail as recommended by the manufacturer. We recommend preparing a 2× stock solution and keeping it at 4 °C. Sodium Deoxycholate, PMSF, Benzamidin, and Protease inhibitor cocktail should be prepared fresh before each experiment.

2.7 Immunoprecipitation Washes and Chromatin Elution

1. RIPA buffer: 50 mM HEPES pH 7.5, 500 mM LiCl, 1 mM EDTA, 1% (v/v) Tergitol® NP40, and 0.7% (w/v) Sodium Deoxycholate.
2. TE buffer: 10 mM Tris–HCl pH 8.0, 1 mM EDTA. Supplemented with 50 mM NaCl.
3. ChIP elution buffer: 50 mM Tris–HCl pH 7.5, 10 mM EDTA, and 1% SDS (w/v).

2.8 Chromatin Reverse-Crosslinking

1. TE buffer: 10 mM Tris–HCl, 1 mM EDTA.
2. RNase cocktail (Ambion) or equivalent mix of RNases (e.g., RNaseA and Rnase T1).

3. 10% SDS in water.
4. 20 mg/mL Proteinase K.
5. DNA purification kit (e.g., Ampure XP beads or similar).
6. Fluorometer for DNA quantification (e.g., Qubit or similar).

2.9 ChIP Enrichment Analysis by qPCR

Standard laboratory reagents required for qPCR quantification, including oligos, qPCR reaction mix, and qPCR instrument.

2.10 Library Preparation

1. High-fidelity PCR master mix: Phusion High-Fidelity PCR Master Mix with HF Buffer (New England Biolabs).
2. PE1 and PE2 10 μM primer mix (Table 1).
3. Optional: Electrophoresis system for DNA size selection (e.g., E-Gel Precast Agarose Electrophoresis System from Thermo Fisher Scientific).
4. Capillar DNA electrophoresis system (e.g., High-sensitivity DNA Bioanalyzer from Agilent or equivalent).

2.11 Illumina Sequencing and Data Analysis

Many bioinformatics tools are available to analyze next-generation sequencing datasets. We recommend using peer-reviewed short read aligners designed for genomics studies and list here a few of these:

1. Short read aligners: Bowtie2 [10], TopHat2 [11], Star [12], BWA [13].
2. Alignment filtering tools: Samtools [14], Picard tools (Java, http://broadinstitute.github.io/picard/), HTSeq (Python) [15], GenomicAlignments (Bioconductor, R) [16].
3. ChIP-seq peak and nucleosome positions callers: MACS [17].
4. Statistical analysis of ChIP-seq datasets: DChIPRep (Bioconductor, R) [18], Csaw (Bioconductor, R) [19], Diffbind (Bioconductor, R) [20].

3 Methods

This protocol has been developed for *Saccharomyces cerevisiae*. We have applied the Bar-ChIP protocol to up to 15 independent samples, and used one-tenth of the obtained pooled chromatin extract (~75 μg of chromatin in total) for each chromatin immunoprecipitation [1]. The protocol can be divided over 3 days as suggested in Fig. 1. All immunoprecipitation steps should be performed in a cold room at 4 °C.

3.1 Crosslink the Chromatin

1. One day before the actual start of the protocol, prepare a liquid pre-culture of *S. cerevisiae* cells from a fresh plate containing rich media for yeast growth (e.g., YPD broth). Using a larger

Table 1
Used oligonucleotides

Oligo	Sequence
PE1.0#	AATGATACGGCGACCACCGAGATCTACACTCTTTCCCTACACGACGCTC TTCCGATC*T
PE2.0#	CAAGCAGAAGACGGCATACGAGATCGGTCTCGGCATTCCTGCTGAACCGCTC TTCCGATC*T
mp1PE1	[Phos]AGCGCTAGATCGGAAGAGCGGTTCAGCAGGAATGCCGAG
mp1PE2	ACACTCTTTCCCTACACGACGCTCTTCCGATCTAGCGCT*T
mp2PE1	[Phos]AAGACTAGATCGGAAGAGCGGTTCAGCAGGAATGCCGAG
mp2PE2	ACACTCTTTCCCTACACGACGCTCTTCCGATCTAGTCTT*T
mp3PE1	[Phos]AAGCGAAGATCGGAAGAGCGGTTCAGCAGGAATGCCGAG
mp3PE2	ACACTCTTTCCCTACACGACGCTCTTCCGATCTTCGCTT*T
mp4PE1	[Phos]ATGTCAAGATCGGAAGAGCGGTTCAGCAGGAATGCCGAG
mp4PE2	ACACTCTTTCCCTACACGACGCTCTTCCGATCTTGACAT*T
mp5PE1	[Phos]ACAGTGAGATCGGAAGAGCGGTTCAGCAGGAATGCCGAG
mp5PE2	ACACTCTTTCCCTACACGACGCTCTTCCGATCTCACTGT*T
mp6PE1	[Phos]ACTCTCAGATCGGAAGAGCGGTTCAGCAGGAATGCCGAG
mp6PE2	ACACTCTTTCCCTACACGACGCTCTTCCGATCTGAGAGT*T
mp7PE1	[Phos]AGCATCAGATCGGAAGAGCGGTTCAGCAGGAATGCCGAG
mp7PE2	ACACTCTTTCCCTACACGACGCTCTTCCGATCTGATGCT*T
mp8PE1	[Phos]AACTTGAGATCGGAAGAGCGGTTCAGCAGGAATGCCGAG
mp8PE2	ACACTCTTTCCCTACACGACGCTCTTCCGATCTCAAGTT*T
mp9PE1	[Phos]AGTTCCAGATCGGAAGAGCGGTTCAGCAGGAATGCCGAG
mp9PE2	ACACTCTTTCCCTACACGACGCTCTTCCGATCTGGAACT*T
mp10PE1	[Phos]ATACGGAGATCGGAAGAGCGGTTCAGCAGGAATGCCGAG
mp10PE2	ACACTCTTTCCCTACACGACGCTCTTCCGATCTCCGTAT*T
mp11PE1	[Phos]ATCTATAGATCGGAAGAGCGGTTCAGCAGGAATGCCGAG
mp11PE2	ACACTCTTTCCCTACACGACGCTCTTCCGATCTATAGAT*T
mp12PE1	[Phos]ATGAGCAGATCGGAAGAGCGGTTCAGCAGGAATGCCGAG
mp12PE2	ACACTCTTTCCCTACACGACGCTCTTCCGATCTGCTCAT*T
mp13PE1	[Phos]CACGATAGATCGGAAGAGCGGTTCAGCAGGAATGCCGAG
mp13PE2	ACACTCTTTCCCTACACGACGCTCTTCCGATCTATCGTG*T
mp14PE1	[Phos]CACTCAAGATCGGAAGAGCGGTTCAGCAGGAATGCCGAG
mp14PE2	ACACTCTTTCCCTACACGACGCTCTTCCGATCTTGAGTG*T

(continued)

Table 1 (continued)

Oligo	Sequence
mp15PE1	[Phos]CAGGCGAGATCGGAAGAGCGGTTCAGCAGGAATGCCGAG
mp15PE2	ACACTCTTTCCCTACACGACGCTCTTCCGATCTCGCCTG*T
mp16PE1	[Phos]CCACGCAGATCGGAAGAGCGGTTCAGCAGGAATGCCGAG
mp16PE2	ACACTCTTTCCCTACACGACGCTCTTCCGATCTGCGTGG*T
mp17PE1	[Phos]CCGCAAAGATCGGAAGAGCGGTTCAGCAGGAATGCCGAG
mp17PE2	ACACTCTTTCCCTACACGACGCTCTTCCGATCTTTGCGG*T
mp18PE1	[Phos]CCTTAGAGATCGGAAGAGCGGTTCAGCAGGAATGCCGAG
mp18PE2	ACACTCTTTCCCTACACGACGCTCTTCCGATCTCTAAGG*T
mp19PE1	[Phos]CGGAATAGATCGGAAGAGCGGTTCAGCAGGAATGCCGAG
mp19PE2	ACACTCTTTCCCTACACGACGCTCTTCCGATCTATTCCG*T
mp20PE1	[Phos]CGTACGAGATCGGAAGAGCGGTTCAGCAGGAATGCCGAG
mp20PE2	ACACTCTTTCCCTACACGACGCTCTTCCGATCTCGTACG*T
mp21PE1	[Phos]CTAGCTAGATCGGAAGAGCGGTTCAGCAGGAATGCCGAG
mp21PE2	ACACTCTTTCCCTACACGACGCTCTTCCGATCTAGCTAG*T
mp22PE1	[Phos]CTATACAGATCGGAAGAGCGGTTCAGCAGGAATGCCGAG
mp22PE2	ACACTCTTTCCCTACACGACGCTCTTCCGATCTGTATAG*T
mp23PE1	[Phos]CTCAGAAGATCGGAAGAGCGGTTCAGCAGGAATGCCGAG
mp23PE2	ACACTCTTTCCCTACACGACGCTCTTCCGATCTTCTGAG*T
mp24PE1	[Phos]CTTGTAAGATCGGAAGAGCGGTTCAGCAGGAATGCCGAG
mp24PE2	ACACTCTTTCCCTACACGACGCTCTTCCGATCTTACAAG*T
mp25PE1	[Phos]GACGGAAGATCGGAAGAGCGGTTCAGCAGGAATGCCGAG
mp25PE2	ACACTCTTTCCCTACACGACGCTCTTCCGATCTTCCGTC*T
mp26PE1	[Phos]GAGTGGAGATCGGAAGAGCGGTTCAGCAGGAATGCCGAG
mp26PE2	ACACTCTTTCCCTACACGACGCTCTTCCGATCTCCACTC*T
mp27PE1	[Phos]GATATAAGATCGGAAGAGCGGTTCAGCAGGAATGCCGAG
mp27PE2	ACACTCTTTCCCTACACGACGCTCTTCCGATCTTATATC*T
mp28PE1	[Phos]GATCAGAGATCGGAAGAGCGGTTCAGCAGGAATGCCGAG
mp28PE2	ACACTCTTTCCCTACACGACGCTCTTCCGATCTCTGATC*T
mp29PE1	[Phos]GCAAGGAGATCGGAAGAGCGGTTCAGCAGGAATGCCGAG
mp29PE2	ACACTCTTTCCCTACACGACGCTCTTCCGATCTCCTTGC*T
mp30PE1	[Phos]GCACTTAGATCGGAAGAGCGGTTCAGCAGGAATGCCGAG

(continued)

Table 1
(continued)

Oligo	Sequence
mp30PE2	ACACTCTTTCCCTACACGACGCTCTTCCGATCTAAGTGC*T
mp31PE1	[Phos]GCTCCAAGATCGGAAGAGCGGTTCAGCAGGAATGCCGAG
mp31PE2	ACACTCTTTCCCTACACGACGCTCTTCCGATCTTGGAGC*T
mp32PE1	[Phos]GGCACAAGATCGGAAGAGCGGTTCAGCAGGAATGCCGAG
mp32PE2	ACACTCTTTCCCTACACGACGCTCTTCCGATCTTGTGCC*T
mp33PE1	[Phos]GGCCTGAGATCGGAAGAGCGGTTCAGCAGGAATGCCGAG
mp33PE2	ACACTCTTTCCCTACACGACGCTCTTCCGATCTCAGGCC*T
mp34PE1	[Phos]GGTAGCAGATCGGAAGAGCGGTTCAGCAGGAATGCCGAG
mp34PE2	ACACTCTTTCCCTACACGACGCTCTTCCGATCTGCTACC*T
mp35PE1	[Phos]GTAGAGAGATCGGAAGAGCGGTTCAGCAGGAATGCCGAG
mp35PE2	ACACTCTTTCCCTACACGACGCTCTTCCGATCTCTCTAC*T
mp36PE1	[Phos]GTGGCCAGATCGGAAGAGCGGTTCAGCAGGAATGCCGAG
mp36PE2	ACACTCTTTCCCTACACGACGCTCTTCCGATCTGGCCAC*T
mp37PE1	[Phos]GTTTCGAGATCGGAAGAGCGGTTCAGCAGGAATGCCGAG
mp37PE2	ACACTCTTTCCCTACACGACGCTCTTCCGATCTCGAAAC*T
mp38PE1	[Phos]TACAGCAGATCGGAAGAGCGGTTCAGCAGGAATGCCGAG
mp38PE2	ACACTCTTTCCCTACACGACGCTCTTCCGATCTGCTGTA*T
mp39PE1	[Phos]TATAATAGATCGGAAGAGCGGTTCAGCAGGAATGCCGAG
mp39PE2	ACACTCTTTCCCTACACGACGCTCTTCCGATCTATTATA*T
mp40PE1	[Phos]TCATTCAGATCGGAAGAGCGGTTCAGCAGGAATGCCGAG
mp40PE2	ACACTCTTTCCCTACACGACGCTCTTCCGATCTGAATGA*T
mp41PE1	[Phos]TCCCGAAGATCGGAAGAGCGGTTCAGCAGGAATGCCGAG
mp41PE2	ACACTCTTTCCCTACACGACGCTCTTCCGATCTTCGGGA*T
mp42PE1	[Phos]TCGGCAAGATCGGAAGAGCGGTTCAGCAGGAATGCCGAG
mp42PE2	ACACTCTTTCCCTACACGACGCTCTTCCGATCTTGCCGA*T

All oligonucleotides are HPLC-purified and produced by Sigma-Aldrich. [Phos] refers to 5′ Phosphorylation; *Refer to S-linkage between the two bases. Indexing barcodes are underlined. Sequences are represented in 5′–3′ orientation. #Oligonucleotide sequences © 2006–2016 Illumina, Inc. All rights reserved vessel, dilute the amount of pre-culture required to reach the exponential phase at the time of harvest and grow cells overnight. Transfer 50 mL of cells in exponential growth phase (OD_{600} = 1 is recommended but may vary depending on the experimental design) into conical tubes (*see* **Note 5**).

2. Add 1.388 mL of formaldehyde (37%) per 50 mL cell culture (~1% final concentration) and incubate at room temperature (20–25 °C) for 15 min on a rocker or orbital shaker (*see* **Notes 1** and **6**).
3. Inactivate crosslinking by adding 2.7 mL 2.5 M glycine and place the tubes back on the shaker for 5 min at room temperature.
4. Harvest the cells by centrifugation (i.e., 2 min at 3500 × *g*) at 4 °C and discard supernatant.
5. Wash cross-linked cell pellets three times with 30 mL of ice-cold TBS. Cells are recovered by centrifugation between each wash. Pellets can be snap-frozen in liquid nitrogen and stored for several weeks at −80 °C.

3.2 Obtain Mono-nucleosomal Particles by Micrococcal Nuclease (MNase) Digestion

1. Resuspend the cell pellets in 0.6 mL ice-cold NP-S buffer supplemented with 24 μL of protease inhibitor cocktail.
2. Transfer the cells to pre-cooled 1.5 mL screw-cap tubes containing ~350 μL glass beads.
3. Break the cells in the cold room by intense bead beating. We used a FastPrep-24 instrument (MP Biomedicals) with four cycles of 20 s at 6.5 m/s and a 1 min pause between each pulse.
4. Briefly spin down the tubes and transfer the supernatant containing broken cell extracts (~600 μL) to 1.5 mL tubes.
5. Add 9 μL of Micrococcal nuclease (~80 U) to each sample, flick tubes and incubate at 37 °C for 45 min with agitation (e.g., agitated thermal block at 900 rpm) (*see* **Note 2**).
6. Place samples on ice, add 12.2 μL EGTA 0.5 M to stop reaction and flick tubes. EGTA will chelate Ca^{2+} ions and inactivate the MNase.
7. Centrifuge samples for 3 min at 6000 × *g* at 4 °C to pellet all cell debris and transfer the supernatant (containing solubilized chromatin and therefore mono-nucleosomes) into new 1.5 mL tubes. This sample will be the Whole Cell Extract (WCE) or Input control.
8. Estimate chromatin concentration by UV-spectrophotometry (A_{260nm}) (e.g., using a Nanodrop instrument).

3.3 Assess MNase Treatment Quality

Before proceeding with the protocol, it is important to assess the quality of the MNase digestion. For many applications, optimal digestion should yield ~80% of mono-nucleosomes.

1. Mix an aliquot of 50 μL of each MNase-treated sample with 200 μL of TE buffer.
2. Add 1 μL of RNase cocktail and incubate for 30 min at 37 °C.
3. Add 15 μL of 10% SDS and of 7.5 μL Proteinase K and incubate for 4 h at 65 °C on a shaking heat block (at ~900 rpm) for reverse-crosslinking.
4. Add 27 μL of 3 M NaOAc (1/10 x sample volume) and 540 μL (2× sample volume) of ice-cold 100% ethanol to each sample. Mix the samples and incubate on ice for 30 min.
5. Precipitate the DNA using a benchtop centrifuge at maximum speed (15,000 × *g*) for 15 min at 4 °C.
6. Remove the supernatant and wash the pellet with 500 μL of ice-cold 70% ethanol. Centrifuge at maximum speed for 5 min.
7. Remove the supernatant and air-dry the pellet for 5 min. Resuspend the pellets in 20 μL of EB buffer. Measure the DNA concentration by UV-spectrophotometry ($A_{260\text{nm}}$).
8. Load between 250 and 500 ng of DNA on a 2% agarose to assess the MNase digestion pattern. Use SYBR Green II or a similar dye to visualize the DNA (*see* **Note** 7 and Fig. 2).

3.4 Antibody Coupling to Magnetic Beads

Use 50 μL of Dynabeads per immunoprecipitation (Ab^+) as well as for the negative control (i.e., without antibody; Ab^-) (*see* **Note 8**).

1. Place the tubes containing the beads on a magnet stand, let the beads settle and remove the supernatant.
2. Wash the beads twice with 600 μL of ice-cold PBS/BSA, removing the supernatant by incubating on the magnetic stand.
3. Resuspend the beads in 30 μL PBS/BSA.
4. For Ab^+ incubation, add 1 μL of yeast tRNA (10 mg/mL) and the amount of Ab as determined by preliminary tests. For instance, 3 μg of the antibody against H3K4me3 from Abcam (# ab8580) was used per IP in our experiments [1]. As previously indicated, the amount of Ab used should be optimized in a case-specific manner. For the Ab^- control, replace the antibody volume with an equivalent volume of PBS/BSA. Incubate the beads with the antibody for 4 h at 4 °C on a rotating wheel to saturate the magnetic beads. Barcoding of the chromatin can be initiated in the meantime (*see* below).

3.5 Chromatin Barcoding

Use 50 μg of chromatin as measured by UV-spectrophotometry for each replicate sample. Samples from the same barcoded chromatin

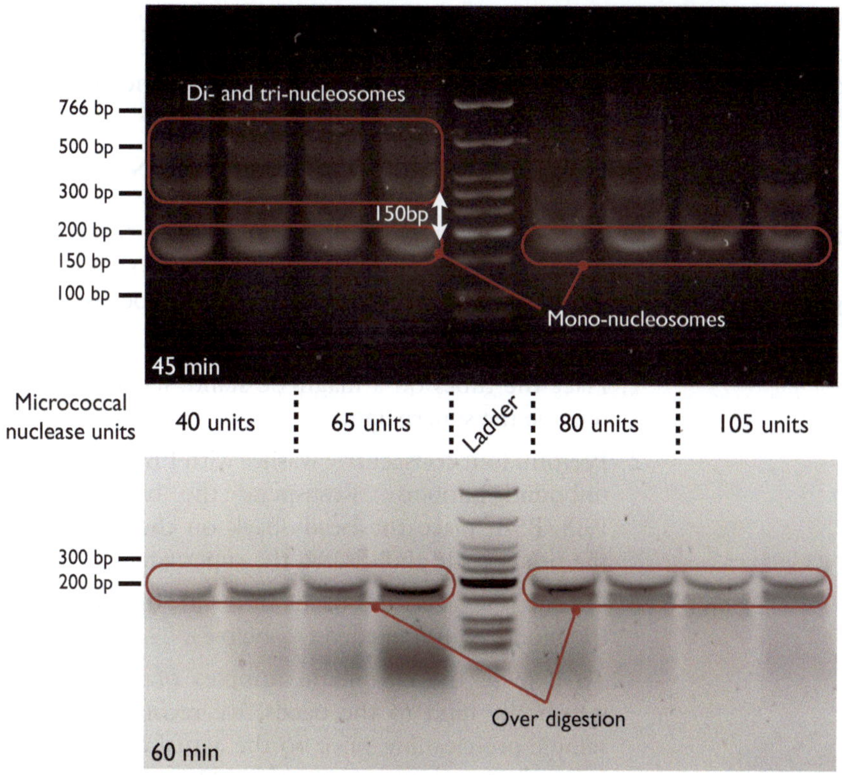

Fig. 2 Selection of the optimal conditions for MNase treatment. MNase digestion was conducted for the equivalent of 50 mL of yeast culture cross-linked at OD ~ 1. Samples were incubated with different amounts of MNase and treatment durations. In the present case, a 45 min-long treatment with 80 units of MNase produces the optimal nucleosome pattern. Samples are visualized using a 2% agarose gel

will be used for the incubation with the antibodies of interest (Ab^+), the negative control immunoprecipitations (Ab^-) and for the bar-coded input (WCE) samples (Fig. 1).

3.5.1 End Repair of the Nucleosome Fragments

1. Increase the volume of each sample (50 μg of chromatin) to 34 μL with water. Add to each sample 4 μL of End Repair Buffer and 2 μL of End Repair Enzyme mix.
2. Incubate the 40 μL reaction mix at 20 °C for 45 min.
3. Heat inactivate at 70 °C for 15 min.

3.5.2 Generate Sticky-Ends by Adding a Protruding dA

1. To each 40 μL of sample from the previous step add: 1 μL of dATP 100 mM, 2 μL of Klenow fragment 3′–5′ exo- and 1 μL of H_2O.
2. Incubate the 44 μL reaction mix at 37 °C for 45 min.
3. Heat inactivate at 70 °C for 5 min.

3.5.3 Ligate Barcoded Adapters for Sample Identification and Library Preparation

1. To each 44 μL of sample from the previous step add: 6 μL of T4 DNA ligase buffer 10×, 4 μL of T4 DNA ligase and 6 μL of barcoded adapter (10 μM) (*see* Table 1 and **Note 4**).
2. Incubate the 60 μL reaction mix at 16 °C for 4 h (*see* **Note 9**).
3. Heat inactivate at 65 °C for 15 min (*see* **Note 10**).

3.6 Chromatin Pre-cleaning and Sample Immunoprecipitation

Wash the antibody-conjugated beads (from Subheading 3.4) to eliminate any unbound antibody. This process should be performed in parallel for both the antibody-coupled beads (Ab^+) and the negative control (Ab^-).

1. Place the tubes on a magnetic stand, let the beads settle and remove the supernatant.
2. Perform four consecutive washes with PBS/BSA to remove any unbound antibody. Resuspend the beads in 600 μL of PBS/BSA, place the beads back on the magnetic stand, let the beads settle and discard the supernatant.
3. Finally, resuspend the beads in 30 μL of PBS/BSA and keep them on ice until needed (*see* below).
4. To deplete the chromatin samples of molecules that non-specifically bind to the beads, we recommend performing a sample pre-cleaning prior to the actual immunoprecipitation. Increase the volume of barcoded MNase-treated samples (from Subheading 3.5) to 400 μL in Lysis buffer at 1× final concentration.
5. In a different tube, pipet 30 μL of Dynabeads (protein A or G) per sample and wash the beads twice with 600 μL of PBS/BSA as previously described.
6. Add the washed beads to the barcoded chromatin sample and incubate them on a rotating wheel for 20 min at 4 °C (in a cold room).
7. After the pre-cleaning step, place the tubes on a magnetic stand and recover the supernatant that contains the pre-cleaned barcoded chromatin. Important: keep this supernatant as it contains the barcoded chromatin fraction.
8. Take a 50 μL aliquot of the pre-cleaned barcoded chromatin and store it at −20 °C (or at 4 °C together with the immunoprecipitating samples). This aliquot will be used to generate the Whole Cell Extract/chromatin Input libraries.
9. Finally, add the 30 μL of washed antibody-coupled beads (or the Ab^- negative control beads) to the remaining pre-cleaned barcoded chromatin. In our hands ~1 mg of chromatin combining up to 15 samples was sufficient for each immunoprecipitation. If multiple antibodies will be used to assay the

same chromatin pool, the pre-cleaned barcoded chromatin can be aliquoted into different tubes.

10. Incubate the barcoded chromatin with the Ab-conjugated beads overnight at 4 °C on a rotating wheel.

3.7 Immuno-precipitation, Washing, and Elution

1. After overnight incubation, place the tubes on a magnetic stand, let the beads settle and remove the supernatant containing the unbound chromatin.
2. Perform eight consecutive washes of the beads with 1 mL RIPA buffer (*see* **Note 11**).
3. Perform one wash with 1 mL TE supplemented with 50 mM NaCl.
4. To elute the sample from the beads, resuspend the beads in 60 μL of ChIP elution buffer and incubate for 10 min at 68 °C in a thermal block with agitation (~900 rpm).
5. Place the tubes on a magnetic rack and immediately transfer 50 μL of the eluate into new 1.5 mL tubes (*see* **Note 12**).
6. To maximize sample recovery, add another 50 μL ChIP elution buffer and incubate for another 10 min at 68 °C with agitation.
7. Recover an additional 50 μL of eluate and combine with the previous eluate from **step 5**.

3.8 Chromatin De-crosslinking

Once samples are recovered, it is necessary to reverse the crosslink so that DNA molecules are accessible to subsequent enzymatic reactions. This process should be applied to both the immunoprecipitated samples and the WCE controls (Input).

1. Increase sample volume to 250 μL with TE buffer.
2. Add 1 μL RNase cocktail to each sample and incubate the sample for 30–60 min at 37 °C.
3. Add 15 μL of 10% SDS and 7.5 μL of Proteinase K and incubate the samples overnight at 65 °C (*see* **Note 13**).
4. Recover the DNA. Use Ampure XP beads according to the manufacturer's instructions, but column or phenol purification are good options too.
5. Add 1.8× volumes of Ampure XP beads to each tube, mix up and down with the pipette and incubate for 5 min at room temperature.
6. Place the sample on a magnetic rack and let the beads settle.
7. Remove the supernatant and perform two washes with 200 μL 70% EtOH without disturbing the beads or removing the tubes from the magnetic rack.
8. Let the beads air-dry for 2 min.
9. Remove the tubes from the magnet and resuspend them in 30 μL EB.

10. After 5 min incubation, place the tubes back on the magnetic rack and recover the supernatant containing the samples of interest.
11. Transfer eluates into new tubes and measure DNA concentration with a fluorometer (e.g., Qubit) using 1 μL per sample.

3.9 Assessment of the Enrichment by qPCR

As the quality and interpretability of any ChIP-seq experiment will depend on the quality of the immunoprecipitation itself, we recommend independently assessing the immunoprecipitation enrichment before sequencing. qPCR analysis should be performed using known binding sites for the protein of interest (if possible) to compare bound regions and negative controls and estimate the immunoprecipitation efficiency. This efficiency should be evaluated and optimized for each antibody before any genome-wide analysis is performed.

3.10 Library Preparation

1. For each library use between 0.5 and 1 ng DNA as template for PCR amplification. Increase the volume to 19.6 μL with water.
2. To each sample, add 20 μL Phusion mix and 0.4 μL PE1/PE2 10 μM primer mix.
3. Perform a PCR with the following conditions: 98 °C 30 s, 18 cycles of (98 °C 30 s, 65 °C 30 s, 72 °C 30 s), final elongation at 72 °C 5 min and hold at 4 °C (*see* **Note 14**).
4. Purify PCR products using Ampure XP beads (1.8× volume) as previously described.
5. Prior to sequencing, the samples may be size-selected by E-Gel or using gel-free methods such as Ampure XP or SPRISelect beads for double-sided size selection as described by the manufacturer. This optional step will allow restricting the size of analyzed fragments (e.g., to focus the sequencing power on mononucleosome particles, and exclude longer reads originating from di- or tri-nucleosome particles).
6. Sample quantity and quality should be assessed by fluorometry (e.g., Qubit) and by High-sensitivity DNA Bioanalyzer, respectively. An example of a successfully amplified library is shown in Fig. 3.

3.11 Illumina Sequencing and Data Analysis

Samples should be submitted for high-throughput sequencing. We recommend using paired-end sequencing as this will provide information on both the position of the DNA-binding particle (center of the reads) and the size of the protected region. This information is critical to exclude any potential di- or tri-nucleosome particles and improve the in silico removal of PCR duplicates. As each sample is identified by a unique barcode, each immunoprecipitation (or WCE control) must be sequenced on an independent Illumina lane. However, this may be circumvented by changing the barcode design and using a dual index system during the PCR amplification

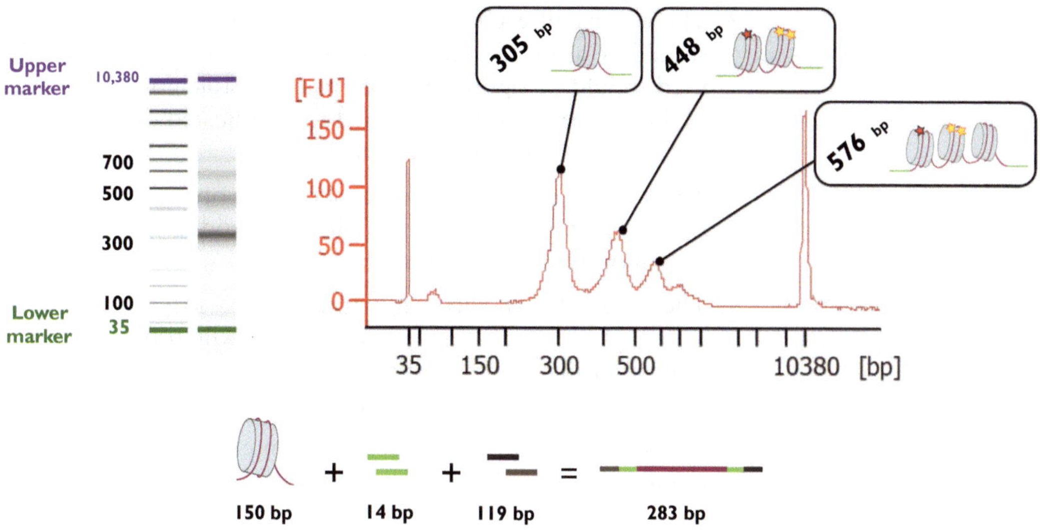

Fig. 3 Example of a successfully amplified library analyzed on a DNA Bioanalyzer (Agilent)

step (that would require further optimization). It is recommended to add up to 10% Illumina PhiX control to the sequencing run to increase complexity, as all reads will contain a T in the seventh base.

Sequencing data can be analyzed similarly to any ChIP-seq approach.

1. Trim the barcodes from the reads and align them to the genome (e.g., using Bowtie2 [10] or some of the aforementioned tools).
2. Filter read pairs with a unique, unambiguous alignment and in convergent orientation (inward-facing read pairs).
3. Use the length of the pairs of reads to assess the particle size and filter out long di-nucleosomes fragments (e.g. 220 bp or longer, this threshold may be adapted based on the high sensitivity DNA Bioanalyzer profiles used to measure the library size).
4. Data analysis and visualization can be performed using available software listed above. For differential analysis of enrichment profiles around genomic regions of interest (transcription start sites or enhancers for example), we recommend using the DChIPRep package (Bioconductor, R) that was developed for this purpose [18].

4 Notes

1. Manipulate with adequate personal protection elements, under a fume-hood and following instructions from the local health and safety offices. Waste containing formaldehyde should be disposed according to the local regulations.

2. The activity of each batch of Micrococcal Nuclease (MNase) should be assessed and the required amount of enzyme to be used should be optimized. The reaction stoichiometry as well as the incubation time may need to be adjusted depending on the type of modifications profiled.
3. The amount of used antibodies should be optimized in a case-specific manner, and their ability to produce good-quality immunoprecipitation should be assayed.
4. The ratio between chromatin and Illumina adapters is critical. Too high a concentration of adapters will produce primer dimers during library amplification. For 50 μg of input chromatin, we recommend using 60 pmol of adapters.
5. In the case of experiments focusing on exponentially growing populations, cells should be exponentially grown for at least three generations (preferentially overnight).
6. We do not recommend going over 15 min of crosslinking as this may affect chromatin fragmentation as well as immunoprecipitation results. Cross-linking times should be optimized in a case-specific manner.
7. We expect to observe an MNase digestion pattern where about 80% of the digested chromatin is in the mononucleosome form and a minority in the dinucleosome or trinucleosome form (*see* Fig. 2).
8. Use the appropriate beads (e.g., Protein G, Protein A, etc.) as recommended by the antibody supplier. The selection of the appropriate antibody is essential to obtain IPs of good quality. Therefore, each antibody lot should be tested and the IP conditions should be optimized for each modification to be profiled. Resources such as the Antibody Validation Database (http://compbio.med.harvard.edu/antibodies/) are extremely useful to preselect good quality antibodies [21].
9. It is critical to perform the ligation reaction for at least 4 h to increase the uniformity of barcoding efficiency across samples and therefore limit sample variability.
10. It is critical to inactivate the ligase to minimize the appearance of nucleosome particles containing unmatched barcodes (due to residual ligase activity after sample pooling).
11. We recommend performing all washes in a 4 °C cold room and using pre-chilled buffers.
12. Avoid disturbing the beads with the pipet tips to minimize bead carryover.
13. The length of time for reverse crosslinking is critical to the subsequent steps of the protocol; shorter incubation times will result in lower efficiency of the following steps, in particular PCR amplification.

14. Several PCR reactions may be run in parallel using 0.5 ng of template and by decreasing the number of cycles to 12–14. PCR reactions can then be pooled together to increase the complexity of the sample to be sequenced. In some cases, the immunoprecipitated sample may be poorly amplified. Several factors may be at play, including incomplete reverse-crossliking or carryover of reagents such as SDS that may decrease the PCR efficiency.

Acknowledgments

The experimental part of this work was performed at EMBL and technically supported by the EMBL Genomics Core Facilities. C. D.C. was supported by a Ph.D. fellowship from the Boehringer Ingelheim Fonds. S.H.A. was supported by an EIPOD Marie Curie/COFUND postdoctoral fellowship. The experimental part of the study was supported by grants from the National Institutes of Health, Deutsche Forschungsgemeinschaft and European Research Council Advanced Investigator Grant (to L.M.S.). Research at VP laboratory is supported by a SciLifeLab Fellowship from the Science for Life Laboratory (SFO Karolinska Institutet), a Starting Grant from the Swedish Research Council (Vetenskapsrådet), a Swedish Foundation's Starting Grant (Ragnar Söderberg Foundation), and a Wallenberg Academy Fellowship (Knut and Alice Wallenberg Foundation). Christophe D. Chabbert and Sophie H. Adjalley contributed equally to this work.

References

1. Chabbert CD, Adjalley SH, Klaus B et al (2015) A high-throughput ChIP-Seq for large-scale chromatin studies. Mol Syst Biol 11(1):777. doi:10.15252/msb.20145776
2. Weiner A, Hsieh TH, Appleboim A et al (2015) High-resolution chromatin dynamics during a yeast stress response. Mol Cell 58(2):371–386. doi:10.1016/j.molcel.2015.02.002
3. Consortium EP (2012) An integrated encyclopedia of DNA elements in the human genome. Nature 489(7414):57–74. doi:10.1038/nature11247
4. Garber M, Yosef N, Goren A et al (2012) A high-throughput chromatin immunoprecipitation approach reveals principles of dynamic gene regulation in mammals. Mol Cell 47(5):810–822. doi:10.1016/j.molcel.2012.07.030
5. Aldridge S, Watt S, Quail MA et al (2013) AHT-ChIP-seq: a completely automated robotic protocol for high-throughput chromatin immunoprecipitation. Genome Biol 14(11):R124. doi:10.1186/gb-2013-14-11-r124
6. Lara-Astiaso D, Weiner A, Lorenzo-Vivas E et al (2014) Immunogenetics. Chromatin state dynamics during blood formation. Science 345(6199):943–949. doi:10.1126/science.1256271
7. van Galen P, Viny AD, Ram O et al (2016) A multiplexed system for quantitative comparisons of chromatin landscapes. Mol Cell 61(1):170–180. doi:10.1016/j.molcel.2015.11.003
8. Weiner A, Lara-Astiaso D, Krupalnik V et al (2016) Co-ChIP enables genome-wide mapping of histone mark co-occurrence at single-molecule resolution. Nat Biotechnol 34(9):953–961. doi:10.1038/nbt.3652
9. Sadeh R, Launer-Wachs R, Wandel H et al (2016) Elucidating combinatorial chromatin states at single-nucleosome resolution. Mol Cell 63(6):1080–1088. doi:10.1016/j.molcel.2016.07.023

10. Langmead B, Salzberg SL (2012) Fast gapped-read alignment with Bowtie 2. Nat Methods 9 (4):357–359. doi:10.1038/nmeth.1923
11. Kim D, Pertea G, Trapnell C et al (2013) TopHat2: accurate alignment of transcriptomes in the presence of insertions, deletions and gene fusions. Genome Biol 14(4):R36. doi:10.1186/gb-2013-14-4-r36
12. Dobin A, Davis CA, Schlesinger F et al (2013) STAR: ultrafast universal RNA-seq aligner. Bioinformatics 29(1):15–21. doi:10.1093/bioinformatics/bts635
13. Li H, Durbin R (2009) Fast and accurate short read alignment with Burrows-Wheeler transform. Bioinformatics 25(14):1754–1760. doi:10.1093/bioinformatics/btp324
14. Li H, Handsaker B, Wysoker A et al (2009) The sequence alignment/map format and SAMtools. Bioinformatics 25 (16):2078–2079. doi:10.1093/bioinformatics/btp352
15. Anders S, Pyl PT, Huber W (2015) HTSeq—a Python framework to work with high-throughput sequencing data. Bioinformatics 31(2):166–169. doi:10.1093/bioinformatics/btu638
16. Lawrence M, Huber W, Pages H et al (2013) Software for computing and annotating genomic ranges. PLoS Comput Biol 9(8): e1003118. doi:10.1371/journal.pcbi.1003118
17. Liu T (2014) Use model-based analysis of ChIP-Seq (MACS) to analyze short reads generated by sequencing protein-DNA interactions in embryonic stem cells. Methods Mol Biol 1150:81–95. doi:10.1007/978-1-4939-0512-6_4
18. Chabbert CD, Steinmetz LM, Klaus B (2016) DChIPRep, an R/bioconductor package for differential enrichment analysis in chromatin studies. PeerJ 4:e1981. doi:10.7717/peerj.1981
19. Lun AT, Smyth GK (2016) csaw: a bioconductor package for differential binding analysis of ChIP-seq data using sliding windows. Nucleic Acids Res 44(5):e45. doi:10.1093/nar/gkv1191
20. Ross-Innes CS, Stark R, Teschendorff AE et al (2012) Differential oestrogen receptor binding is associated with clinical outcome in breast cancer. Nature 481(7381):389–393. doi:10.1038/nature10730
21. Egelhofer TA, Minoda A, Klugman S et al (2011) An assessment of histone-modification antibody quality. Nat Struct Mol Biol 18 (1):91–93. doi:10.1038/nsmb.1972

Chapter 17

Analysis of ChIP-seq Data in R/Bioconductor

Ines de Santiago and Thomas Carroll

Abstract

The development of novel high-throughput sequencing methods for ChIP (chromatin immunoprecipitation) has provided a very powerful tool to study gene regulation in multiple conditions at unprecedented resolution and scale. Proactive quality-control and appropriate data analysis techniques are of critical importance to extract the most meaningful results from the data. Over the last years, an array of R/ Bioconductor tools has been developed allowing researchers to process and analyze ChIP-seq data. This chapter provides an overview of the methods available to analyze ChIP-seq data based primarily on software packages from the open-source Bioconductor project. Protocols described in this chapter cover basic steps including data alignment, peak calling, quality control and data visualization, as well as more complex methods such as the identification of differentially bound regions and functional analyses to annotate regulatory regions. The steps in the data analysis process were demonstrated on publicly available data sets and will serve as a demonstration of the computational procedures routinely used for the analysis of ChIP-seq data in R/Bioconductor, from which readers can construct their own analysis pipelines.

Key words ChIP-seq, Bioconductor, Workflow, Sequencing, Computational analysis, Data analysis

1 Introduction

ChIP-seq (Chromatin immunoprecipitation followed by high-throughput sequencing) technology has revealed unprecedented insights and discoveries into gene regulation events in various diseases and biological pathways, such as cancer and development. ChIP-seq allows researchers to study the landscape of chromatin modifications or the binding patterns of transcription-factors or other chromatin-associated proteins like RNA-polymerases or chromatin modulators. The success of a ChIP-seq experiment is largely dependent on the final signal-to-noise ratio (i.e., the level of enrichment over non-specifically precipitated DNA). Several factors have been shown to affect the quality of the results [1, 2], from experimental factors, such as the expression level of the protein of interest in the assayed tissue or the efficiency of the antibody, to technical factors inherent to the ChIP-seq technology. The large datasets generated, the extensive quality issues, and the diversity of

Neus Visa and Antonio Jordán-Pla (eds.), *Chromatin Immunoprecipitation: Methods and Protocols*,
Methods in Molecular Biology, vol. 1689, DOI 10.1007/978-1-4939-7380-4_17,

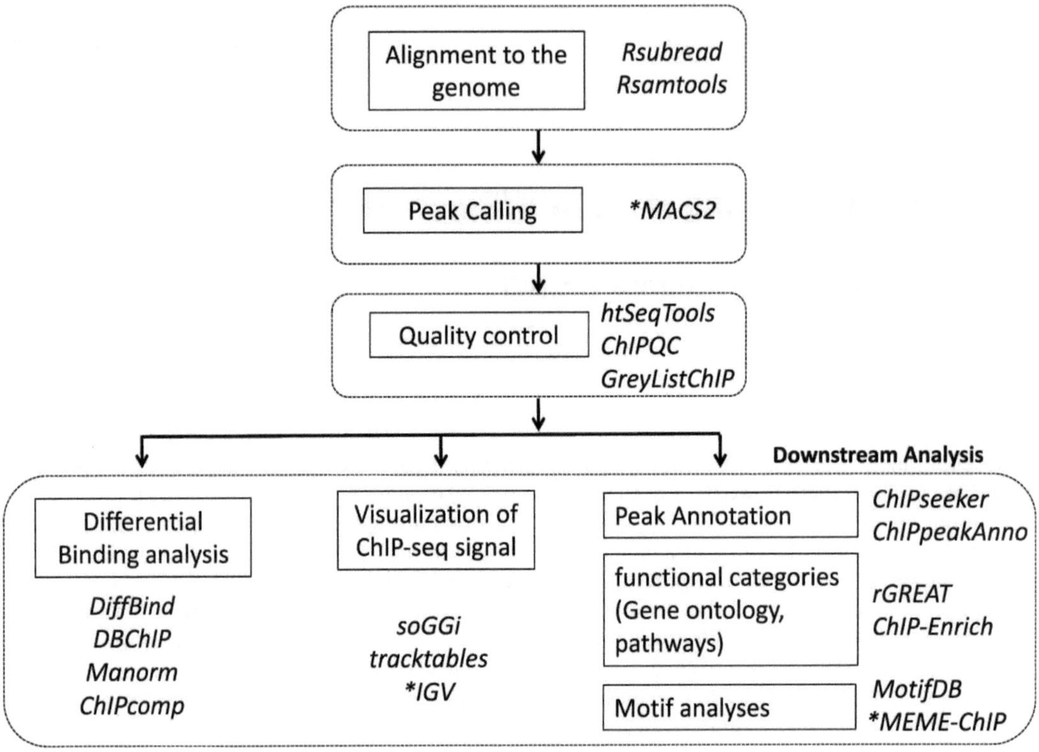

Fig. 1 ChIP-seq analysis workflow. *Boxes* indicate the steps involved in ChIP-seq workflow for various aims discussed in this protocol, with examples for different generally available R software packages (*asterisk* some example tools are not R packages but are widely used and so were included)

ChIP-seq data have turned its bioinformatic analysis into a challenging task. ChIP-seq data analysis involves many computational analysis steps (Fig. 1), a robust computational protocol is necessary for extensive quality control steps to identify artifacts introduced during sample preparation, or technical biases introduced during sequencing. Over the recent years, several R/Bioconductor tools have been developed, allowing researchers to process and analyze ChIP-seq data. In this chapter, we provide an overview of the methods available and describe a computational protocol for the analyses of ChIP-seq data, based primarily on software packages from the open-source Bioconductor project.

We will group the computational analysis steps according to the tasks they address:

1. Data pre-processing and alignment. This is addressed by methods that map reads to the genome and perform basic sorting and indexing of BAM files.
2. Peak calling. The genomic regions that are significantly enriched for sequencing reads from ChIP material, compared with reads from control or input material, are regarded as peaks. Called peaks represent candidate regions of targeted

protein/DNA-binding and histone modification sites. Other genomic regions are regarded as "non-specific background" or "noise."

3. ChIP-seq quality control. This is addressed by various quality control checks and procedures to ensure that technical effects do not distort downstream results.
4. Identification of differential binding. These include methods that compare relative depth of coverage or peak presence across two or more conditions.
5. Data visualization through average profiles and coverage tracks. An overview of tools that facilitate the visualization of the ChIP-seq data from R.
6. Functional downstream analysis. This is addressed by methods that include determination of target genes, enrichment analysis (gene ontology and pathway analysis), and motif finding.

2 Materials

2.1 Experimental Data

ChIP-seq data sets used in this chapter were obtained from ENCODE/Stanford/Yale mouse ChIP-seq experiments in the ENCODE project (GSE36030) [3]. These datasets were generated and analyzed by the labs of Michael Snyder at Stanford University. The data correspond to ChIP-seq data for c-Myc transcription factor, obtained in murine MEL (leukemia) and CH12 (B-cell lymphoma) cell lines. The exact data can be found at the SRA archive or Gene Expression Omnibus under the following accession numbers:

- c-Myc tag dataset obtained in Ch12 cell lines (two replicates). GEO: GSM912906. SRA: SRX140369.
- Ch12 input dataset (one replicates). GEO: GSM912904. SRA: SRX140397.
- c-Myc tag dataset obtained in MEL cell lines (two replicates). GEO: GSM912934. SRA: SRX140367.
- MEL input dataset (one replicates). GEO: GSM912918. SRA: SRX140381.

2.2 Reference Datasets for Alignment

To align sequence reads from ChIP-seq data, an organism's reference genome sequence must be acquired. We used the mouse mm10 genome (Genome Reference Consortium Mouse Genome Assembly GRCm38.p5). The mm10 genome was downloaded in FASTA format as separate chromosomes from NCBI ftp site [4].

2.3 Reference Datasets for Quality-Control

A catalog of known ChIP-seq specific artifact regions for the mouse genome were generated by Alan Boyle and Anshul Kundaje as part of the ENCODE and modENCODE projects [3].

2.4 System Requirements

2.4.1 Hardware Specifications

The entire ChIP-seq computational protocol was run on a MacBook Pro laptop with 16 GB RAM, 3.1 GHz Intel Core i7 and 2 cores. The workflow processes 25 GB of raw FASTQ files, originates 2 GB of results, and takes 193 min (3 h and 13 min) to complete. A typical Linux workstation with 64 GB and 6 cores completes it in 77 min (1 h and 17 min). We note that larger ChIP-seq projects will require a larger amount of RAM and disk space to complete some downstream analysis tasks. We recommend the following hardware for performing ChIP-seq data analysis:

We recommend the following hardware for performing ChIP-seq data analysis:

- For pre-processing (read alignment and peak calling; Fig. 1) and downstream analysis, we recommend a minimum of a Linux/ MacOS 64-bit computer with 16 GB RAM and 2 cores. For every 2 additional cores, a minimum of 8 GB of RAM should be included.
- Downstream analysis (Fig. 1) can be accomplished with a smaller set up. A typical desktop computer (e.g. Intel i5/i7/ Xeon, 4 cores), for example, a PC computer running Windows 8, 10, Server 2008, Server 2012 or a 64-bit MAC computer (operating system version > 10.10.5 is recommended) with 16 GB of RAM.

2.4.2 Required Software and Bioconductor Packages

The following software should be installed on your machine:

1. Most recent version of R (http://www.r-project.org/). Recommended version R > 3.3. Additionally, we recommend R Studio desktop (http://www.rstudio.com/) for data analysis in R.
2. The latest version of Bioconductor. This can be installed by running the command:

```
source("https://bioconductor.org/biocLite.R")
biocLite()
```

3. A robust text editor. We recommend notepad++ for Windows/PC (http://notepad-plus-plus.org/), gEdit for Linux (http://projects.gnome.org/gedit/) and TextWrangler for MAC (http://www.barebones.com/products/textwrangler/download.html).
4. An internet browser with Java support such as Firefox or Safari.
5. Integrative Genomics Viewer (IGV) version > 2.3 (http://www.broadinstitute.org/igv/).

6. Java—The visualization program IGV requires Java. If you do not have it installed (https://java.com/en/download/).
7. Macs2—The Macs2 peak calling program (https://github.com/taoliu/MACS).

We recommend using the most updated available versions of each Bioconductor package at the time of analysis. The Bioconductor packages needed to run this protocol are listed below (the versions used at the time of writing this protocol are included in brackets):

- SRAdb (v1.32), Biostrings (v2.42.1), rtracklayer (v1.34.2), R.utils (v2.5.0), Rsubread (v1.24.1), BiocParallel (1.8.1), Rsamtools (v1.26.1), ChIPQC (v1.10.2), DiffBind (v2.2.8), soGGi (v.1.4.6), GenomicAlignments (v1.10.0), GenomicRanges (v1.26.3), tracktables (v1.8.1), tools (v3.3.3), BSgenome.Mmusculus.UCSC.mm10 (v1.4.0), TxDb.Mmusculus.UCSC.mm10.knownGene (v3.4.0), org.Mm.eg.db (v3.4.0), GenomeInfoDb (v1.10.3), ChIPseeker (v1.10.3), rGREAT (v1.6.0), ChIPpeakAnno (v3.8.9), MotifDb (v1.16.0), and seqLogo (v1.40.0).

To install all necessary Bioconductor packages on your machine run the function *biocLite*. For instance, to install SRAdb, run:

```
biocLite ("SRAdb")
library(SRAdb)
```

3 Methods

3.1 Gathering Publicly Available Data from the Sequence Read Archive (SRA)

3.1.1 Downloading ChIP-Seq Example Datasets

Following publication or release of a study, it is common practice to deposit any associated high-throughput sequencing (HTS) data in a publicly available data repository. Popular repositories for HTS data include those repositories historically associated with microarray data such as Gene Expression Omnibus (GEO) and ArrayExpress as well as members of the International Nucleotide Sequence Database Collaboration (INSDC) including the Short Read Archive (SRA), European Nucleotide Archive (ENA), and the DDBJ Sequence Read Archive (DRA).

In this example, we will query the sequence data centric INSDC repositories making use of the SRAdb Bioconductor library to download raw sequencing reads. We retrieve the data as gzipped FASTQ to avoid the conversion steps required to make use of data in NCBI's own SRA format.

In this code section, we load the SRAdb [5] library and download a local repository of samples' metadata and HTS data ftp locations (SRAmetadb.sqlite) using the *dbConnect* function. Following the acquisition of the local SQL database, we can directly query and retrieve by the samples' SRX IDs using the *getSRAfile* function and specifying the desired file type, location, and download method.

```
library(SRAdb)
sqlfile <<- getSRAdbFile()
sra_con <- dbConnect(SQLite(),sqlfile)

allFilesNeeded <- c("SRX140369","SRX140397","SRX140367", "SRX140381")
sampleGroups <- c("Ch12_cMyc","Mel_cMyc",
                  "Mel_Input_Rep1",
                  "Ch12_Input_Rep1")
dir.create("fastq",showWarnings=FALSE)

getSRAfile(allFilesNeeded, sra_con,
           destDir=file.path(getwd(),"fastq") , fileType='fastq',
           srcType='ftp', makeDirectory=FALSE,
           method='curl', ascpCMD=NULL )
```

3.1.2 Downloading Reference Genome mm10 Datasets

To work with HTS from ChIP-seq experiments, we will first need to retrieve the required reference data. Here, we will work with the mouse mm10 genome (Genome Reference Consortium GRCm38). To align the HTS data, we will need the full sequence information for mm10 in FASTA format.

In the code section below, we download the mm10 genome in FASTA format as separate chromosomes from NCBI ftp. To construct a single FASTA file containing all chromosomes, we use the Biostrings [6] library reading in separate chromosomes with the *readDNAStringSet* function, renaming chromosomes, and writing to a single FASTA file using the *writeXStringSet* function.

```
ftp <- "ftp://ftp.ncbi.nih.gov/genomes/genbank/vertebrate_mammalian/Mus_musculus/
all_assembly_versions/GCA_000001635.7_GRCm38.p5/GCA_000001635.7_GRCm38.p5_assem-
bly_structure/Primary_Assembly/assembled_chromosomes/FASTA"

fastaFileFTP <- c(file.path(ftp,paste0("chr", 1:19, ".fna.gz")),
                  file.path(ftp,paste0("chr", c("X","Y"), ".fna.gz")))
dir.create("mm10_Fasta")

lapply(fastaFileFTP, function(x) {
  download.file(x, file.path("mm10_Fasta", basename(x)))})

library(Biostrings)
fastaFileLocal <- dir("mm10_Fasta", pattern="*\\.fna.gz", full.names=T)
mm10GenomeFasta <- readDNAStringSet(fastaFileLocal)
names(mm10GenomeFasta) <- gsub("\\.fna\\.gz","",
                               basename(fastaFileLocal))
writeXStringSet(mm10GenomeFasta, file.path("mm10_Fasta", "All.fa"))
```

Now that we have the genome sequence in a suitable format, we load the Rsubread [7] package to create the index required for sequence alignment using the *buildindex* function.

```
library(Rsubread)
buildindex("mm10_Fasta/mm10", file.path("mm10_Fasta","All.fa"))
```

3.1.3 Download Data for QC

In addition to the reference sequence required for alignment, we will download a catalog of known ChIP-seq specific artifact regions for the mouse genome. Following work on ChIP-seq generated within the Encode project [3], regions of ultra-high signal ubiquitous to ChIP and input HTS data, known as blacklists, were identified for the Human genome. Further to the work in Humans and from aggregate analysis of ChIP-seq and input data within many model organisms by Alan Boyle and Anshul Kundaje, sets of blacklists for human (hg19), mouse (mm9), *Drosophila* (dm3), and worm (ce10Worm) have been released. Since we will be working with the mm10 genome, we download the liftOver chain for mm9 to mm10 from UCSC which will allow us to map positions within the mm9 genome to mm10. Having downloaded the mm9 Blacklist and liftOver chain, we can make use of the *liftOver* function in the rtracklayer [8] package to update Blacklist's coordinates and create the mm10 Blacklist. Custom blacklists for genomes may also be created by the GreyListChIP [9] Bioconductor package from input samples in the case of no genome blacklist being available or the presence of additional artifact regions found in some cell lines.

```
library(rtracklayer)
download.file("https://personal.broadinstitute.org/anshul/projects/mouse/blacklist/
mm9-blacklist.bed.gz","mm9-blacklist.bed.gz")
mm9BlackList <- import.bed("mm9-blacklist.bed.gz")
download.file("http://hgdownload.cse.ucsc.edu/goldenPath/mm9/liftOver/mm9ToMm10.
over.chain.gz","mm9ToMm10.over.chain.gz")

library(R.utils)
gunzip("mm9ToMm10.over.chain.gz")
chain <- import.chain("mm9ToMm10.over.chain")
mm10BlackList <- liftOver(mm9BlackList, chain)
export.bed(unlist(mm10BlackList), con="mm10-blacklist.bed")
```

3.2 Data Alignment

3.2.1 ChIP-Seq Alignment in R

Now we have retrieved and created all the required reference files for mm10 mouse genome as well as the raw ChIP-seq and input HTS data for the samples of interest.

To align the sequence data for this chapter, we use the Rsubread [7] Bioconductor package. The Rsubread package wraps the standalone Subread aligner, providing an R interface to DNA and

RNA alignment using a seed-and-vote method as well as SNP calling. Similar wrappers are available for other popular aligners including the Bowtie2 and SpliceMap wrappers, RBowtie [10] and QuasR [11].

In this code segment, we load the Rsubread package and align the data using the *align* function specifying the directory for the index we built, FASTQ files, type of alignment, and the names of the output files. Here, we also specify the number of threads to be used in read mapping which reduces overall time of alignment while increasing CPU and memory loads. By default, the *align* function in the Rsubread package omits reads which map to multiple positions in the genome. This will influence the quantification of reads mapping to repeats and so must be considered in the interpretation of ChIP quality metrics and global ChIP signal distribution (*see* **Note 1**). The removal of multimapping reads prior to peak calling is recommended to avoid the inclusion of artifacts in identified putative binding sites (*see* **Note 2**).

```
library(Rsubread)
dir.create("bam")
numParallelJobs <- 4
fastqFiles <- dir(pattern="fastq.gz", full.names=TRUE, recursive=TRUE)
bamNames <- basename(gsub("fastq\cr.gz","bam", fastqFiles))
align(index="mm10_Fasta/mm10",
      fastqFiles,
      type="dna",
      output_file=file.path("bam",bamNames),
      nthreads=numParallelJobs)
```

3.2.2 Sorting and Indexing BAM Files with Rsamtools

Once we have aligned the data, we coordinate sort and index the aligned reads in BAM format. BAM format offers greater data compression than standard text formats and, when sorted and indexed, allows for the random access of aligned data through the specification of genomic regions. Here, we use the Rsamtools package [12] and the *sortBam* and *indexBam* functions to sort and index, respectively.

```
library(BiocParallel)
register(BiocParallel::MulticoreParam(workers=numParallel Jobs))
library(tools)
library(Rsamtools)

bplapply(1:length(bamNames), function(x) {
  sortBam(file.path("bam", bamNames[x]),
         file.path("bam", paste("Sorted",
                  file_path_sans_ext(bamNames[x]), sep="_")))
})

indexBam(file.path("bam", paste("Sorted_", bamNames, sep="")))
```

3.3 Peak Calling

A common goal of ChIP-seq experiments is to identify and quantify epigenetic events such as regions of the genome bound by a transcription factor (TF) or those bearing an epigenetic mark (histone modifications). Many strategies have been taken to identify such regions from ChIP-seq data with such strategies often subdivided into the event type under interrogation. TF ChIP-seq will often generate narrow, punctuate sources of high signal whereas some epigenetic marks such as H3K36me3 and H3K27me3 may cover 100's to 10000's of bases with diffuse signal [13] and so the need for specialized software or software modes to identify these distinct patterns of enrichment is apparent.

The identification of epigenetic events from ChIP-seq and ChIP-chip data is broadly termed "peak calling" although the use of this phrase is best suited to the punctuate signal seen in transcription factor ChIP-seq as with the c-Myc ChIP-seq used in this protocol.

Peak calling software for TFs may exploit several features of their ChIP-seq signal enrichment. Broadly, the two most common strategies for identifying TF binding are using ChIP-seq signal enrichment over input sample signal (or local ChIP background when input is unavailable) such as the popular MACS/MACS2 [14] or interrogating the distribution of reads and shape of coverage around binding events as seen in SPP [15], TPIC [16], and Triform [17] peak callers.

In this example, we use the MACS2 peak calling software to allow us to identify both the punctuate and broader regions of enrichment commonly seen for Myc ChIP-seq (*see* **Note 3**). MACS2 first assesses the strand shift showing the highest correlation between Watson and Crick reads to estimate the fragment length and construct the predicted full ChIP signal following reads' extension to their predicted fragments. Following construction of predicted ChIP-signal, MACS2 identifies windows showing enrichment over the local background or that of the input signal. Although MACS2 may be used to call peaks without an input sample, for greater sensitivity it is recommended to use an appropriate input sample for peak calling. Here, we make use of the input sample data for the appropriate cell-lines.

In the code example, we have wrapped a system call to the MACS2 python package to allow us to specify the input and output directories from within R. Here, we make a system call to "Macs2 –callpeaks" command providing the test sample and control samples for the respective cell lines and the directory and name for output files.

```
bamNames <- paste0("Sorted_",
                   basename(gsub("fastq\cr.gz","bam",fastqFiles)))

testSampleBAMs <- file.path("bam",
                            c(bamNames[5], bamNames[6],
                              bamNames[2], bamNames[3]))

testSampleControls <- file.path("bam",
                                c(bamNames[1], bamNames[1],
                                  bamNames[4], bamNames[4]))

namesForMACS2 <- basename(file_path_sans_ext(testSampleBAMs))

for(i in 1:length(namesForMACS2)){
  system(paste0("macs2 callpeak -t ", testSampleBAMs[i],
                " -c ", testSampleControls[i],
                " -name ", namesForMACS2[i],
                " -outdir ","peaks"))}

allBams <- c(testSampleBAMs, unique(testSampleControls))
allPeaks <- c(file.path("peaks",
                        paste0(namesForMACS2, "_peaks.xls")),rep(NA,2))
```

3.4 ChIP-Seq Quality Control

To best characterize the epigenetic events under investigation by ChIP-seq, it is essential to evaluate the quality of the ChIP sample and input control data sets [1, 2, 18]. Several metrics specific to ChIP-seq quality have been proposed including the Encode consortium's use of cross-correlation and the fraction of reads in peaks [3], metrics such as Standardized Standard Deviation and Gini Coefficient of genome coverage available in the htSeqTools R package [19] and Cistrome's use of distribution of signal over peaks and across genomic features [20].

The ChIPQC [18] Bioconductor package calculates selected non-redundant, ChIP-seq-specific quality metrics alongside the appropriate data filtering. In this example, we construct ChIPQC-sample objects for each sample and input in our experiment using the *ChIPQCsample* function while providing the mm10 Blacklist generated earlier and the genome of interest. ChIPQC can gather these metrics across a subset of chromosomes and so here we gather metrics across chromosomes 1–10. Having created a named list of ChIPQCsample objects, we can assess the quality metrics generated within ChIPQC with its plotting functions.

```
register(BiocParallel::MulticoreParam(workers=numParallel Jobs))
library(ChIPQC)
library(rtracklayer)
mm10Annotation <- ChIPQC::getAnnotation("mm10", NULL)
blacklist <- import.bed("mm10-blacklist.bed")
library(rtracklayer)

chipqcList <- bplapply(1:length(allBams),
                    function(x) {
                      ChIPQCsample(allBams[x],
                                   allPeaks[x],
                                   blacklist=blacklist,
                                   annotation=mm10Annotation,
                                   chromosomes=paste0("chr",
                                                      1:10))})

names(chipqcList) <- c("MEL_1","MEL_2", "Ch12_1",
                    "Ch12_2","Input_MEL","Input_Ch12")
```

3.4.1 Cross-correlation and Cross-coverage

Cross-correlation and cross-coverage provide a method of characterizing both the clustering of Watson and Crick reads around the centers of binding sites and the background signal coming from artifact regions such as the Blacklists. ChIPQC uses cross-coverage to identify both a fragment length peak and a read length peak in cross-coverage score. The position and magnitude of any peak away from the read length peak indicates both the fragment length itself as well as the global strength of ChIP signal. The height of the read length peak has been shown to be correlated with the extend of artifact signal, such as seen in Blacklists [18].

In the below example, we plot the cross-coverage for Myc samples and input data using the *plotCC* function and specifying to facet and color by "Sample" using the *facetBy* and *colourBy* parameter.

The cross-coverage profiles shown in Fig. 2 for MEL Myc ChIP-seq samples show that both samples have similar fragment lengths and strengths of ChIP-signal. The read length peaks for MEL however show a stronger CC score at read length for the second replicate (MEL_2) indicating a stronger artifact signal within this replicate. The two Ch12 replicates (Ch12_1 and Ch12_2) show distinct fragment length signal heights and fragment length peak positions highlighting potential sources of variation between these replicates. The CC score at read length position in the input samples is closer to that at the fragment length indicating, as expected, a higher representation of signal within artifact regions in input samples.

```
plotCC(chipqcList, facetBy="Sample", colourBy="Sample")
```

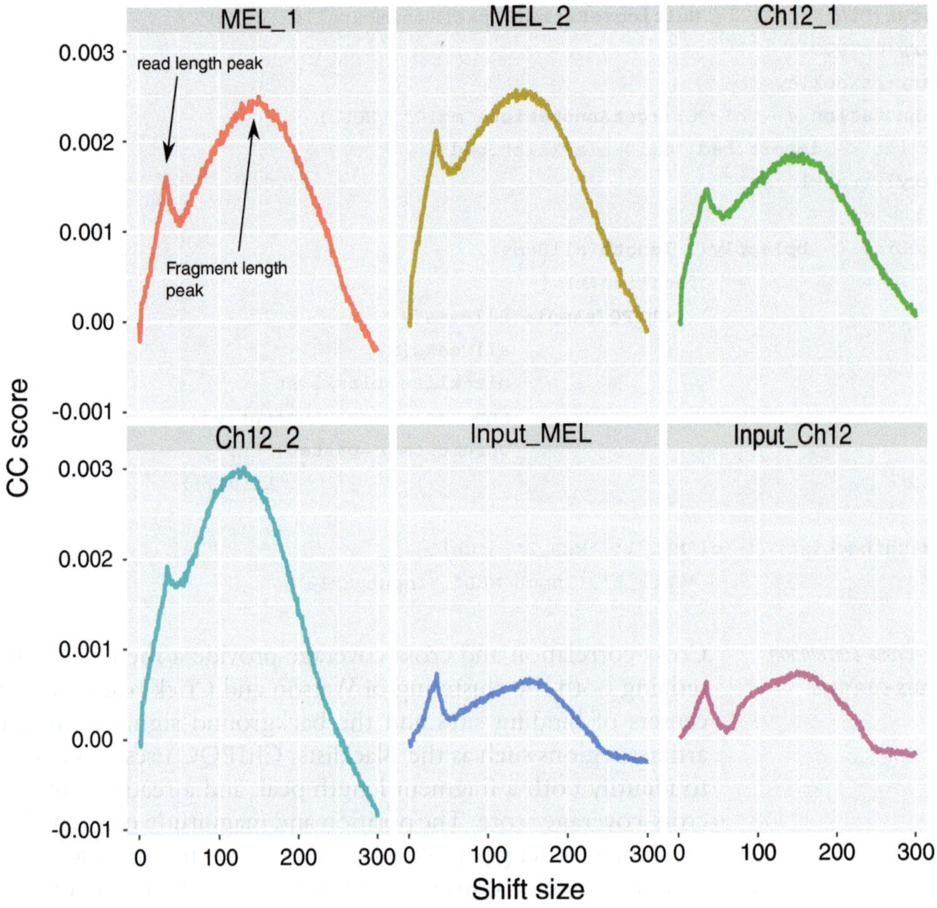

Fig. 2 Example ChIPQC cross-coverage plot grouped by sample ID

3.4.2 Standardized Standard Deviation (SSD)

The Standardized Standard Deviation (SSD) metric provides a measure of inequality of coverage across the genome. SSD however is very sensitive to the signal originating from blacklisted regions [18] and so ChIPQC assess SSD both before and after blacklisting. By plotting SSD before and after blacklisting, ChIPQC provides both a measure of the signal coming from blacklisted regions and those from the regions outside of artifacts.

Here, we use the *plotSSD* function to produce the pre- and post-blacklisting SSD scores shown in Fig. 3a, b. It is clear from the SSD plot that the control samples have the highest scores prior to Blacklisting indicating the expected strong acquisition of signal in artifact regions for input samples. Within the Myc ChIP-seq samples, we see that both Ch12_1 and MEL_2 have higher pre-blacklisting SSD scores matching the higher CC score at read length peaks seen for these samples in the cross-coverage plots of Fig. 2.

For the post-blacklisting scores, we can see that the control samples have the lowest SSD score after blacklisting (Fig. 3b), in

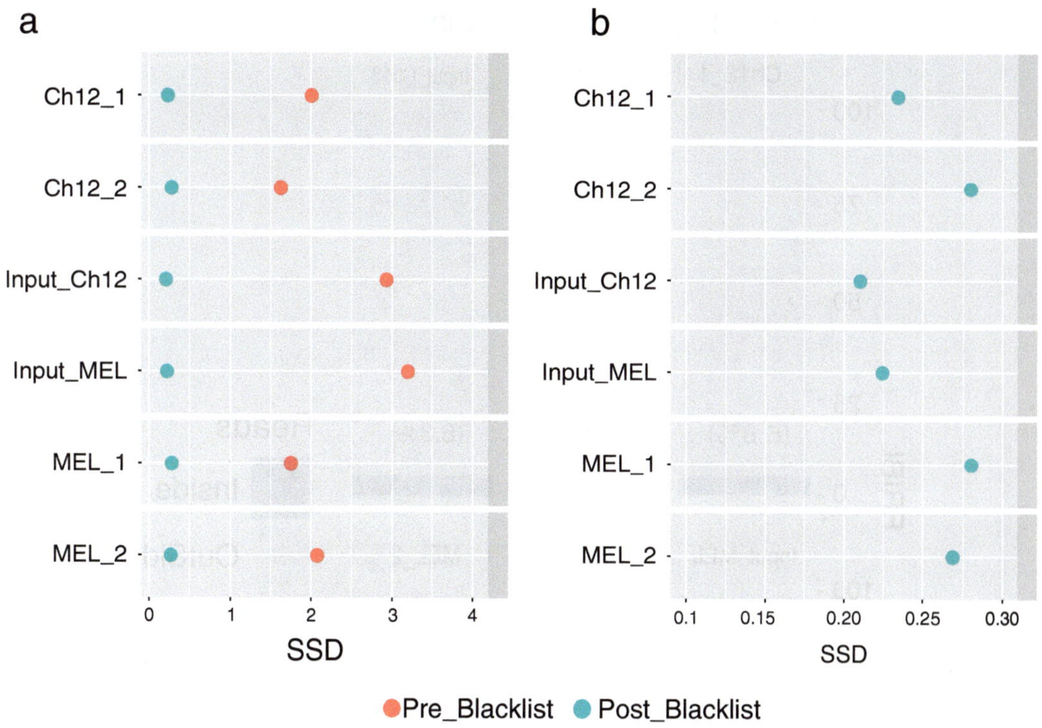

Fig. 3 Example ChIPQC plot of pre- and post-blacklisting SSD scores: (**a**) SSD plot, (**b**) Zoomed SSD plot to better visualize the differences in the post-blacklisting scores

keeping with the majority of background signal being in Blacklist regions. The c-Myc samples all show similar post-blacklisting scores except for the Ch12_2 sample which also showed the highest fragment length peak in CC plot.

```
plotSSD(chipqcList)
plotSSD(chipqcList)+xlim(0.1,0.3)
```

3.4.3 FRiBL (Fraction of Reads in BlackList) and FRiP (Fraction of Reads in Peaks)

In the next step, we plot the percentage of ChIP-seq reads landing in Blacklists or within called peaks using the ChIPQC *plotFribl* and *plotFrip* functions respectively.

The Fribl plots in Fig. 4 show an expected high proportion of signal in Blacklists for input samples as well as Ch12_1 and MEL_2 showing higher background acquisition than their replicates. The Frip plots in Fig. 5 show a similar pattern of scores to that seen with SSD post-blacklisting with the Ch12_1 replicate showing a significantly lower score than other ChIP samples. This highlights the usefulness of SSD filtering where post-blacklisted scores have predicted ChIP-seq signal without the requirement for a peak calling step.

```
plotFribl(chipqcList)
plotFrip(chipqcList)
```

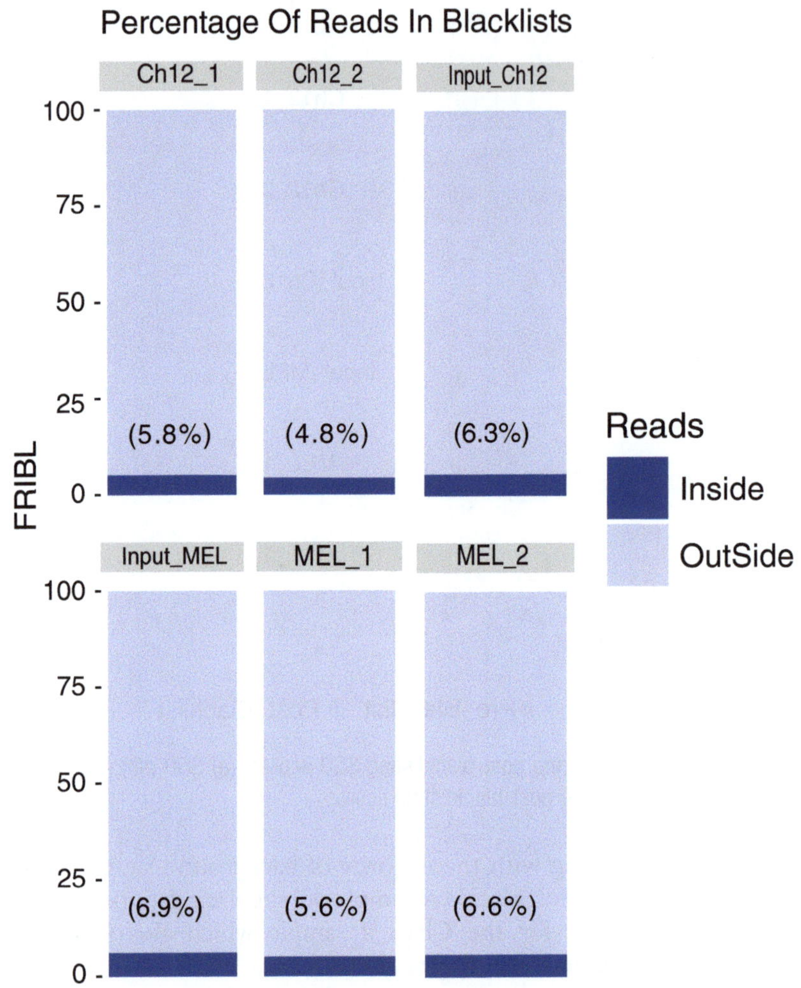

Fig. 4 Example ChIPQC plot showing reads overlapping Blacklists. *Numbers in brackets* indicate the percentage of reads in blacklists for each ChIP and input sample

3.4.4 Distribution of ChIP-Seq Signal Over Genomic Regions

ChIPQC also allows for the assessment of the distribution of ChIP-seq signal over genomic regions. By default, these may be components of genes (5′/3′UTRs, TSSs, Introns, Exons) but user-defined regions may be included. By assessing the distribution across samples, unexpected patterns may be identified in ChIP replicates and within input/control samples.

In the example seen in Fig. 6, we plot the relative enrichment within samples using the plotRegi function. Here, we can observe the expected relative absence of enrichment within or across genes for the control samples. We can also see that the MEL_2 and Ch12_1 samples show lower enrichment within and around the TSS region, correlated with their lower scores for other metrics measuring ChIP-seq signal strength.

```
plotRegi(chipqcList)
```

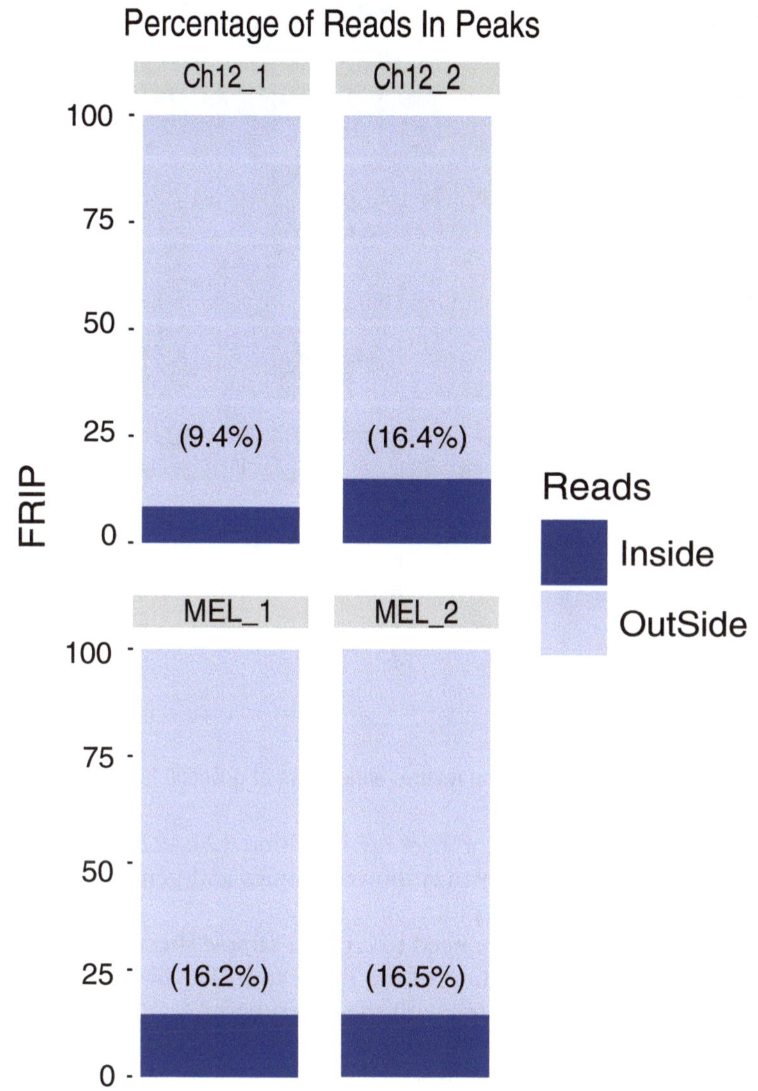

Fig. 5 Example ChIPQC plot showing reads overlapping peaks plot. *Numbers in brackets* indicate the percentage of reads in peaks for each ChIP sample

3.5 Identification of Differential Binding

After obtaining peak sets, a typical research question is to find peak similarities and differences among sample groups (*see* **Notes 4** and **5**). The simplest approach is to perform a so-called "occupancy analysis." The occupancy analysis identifies common and unique peaks by overlapping peak sets between sample groups. Another, often preferred, type of analysis is the "affinity analysis." An affinity analysis quantitatively tests differences in binding affinity between sample groups by testing differences in the intensities of binding profiles, which necessitates read normalization. The process to perform a differential binding analysis is encapsulated in the

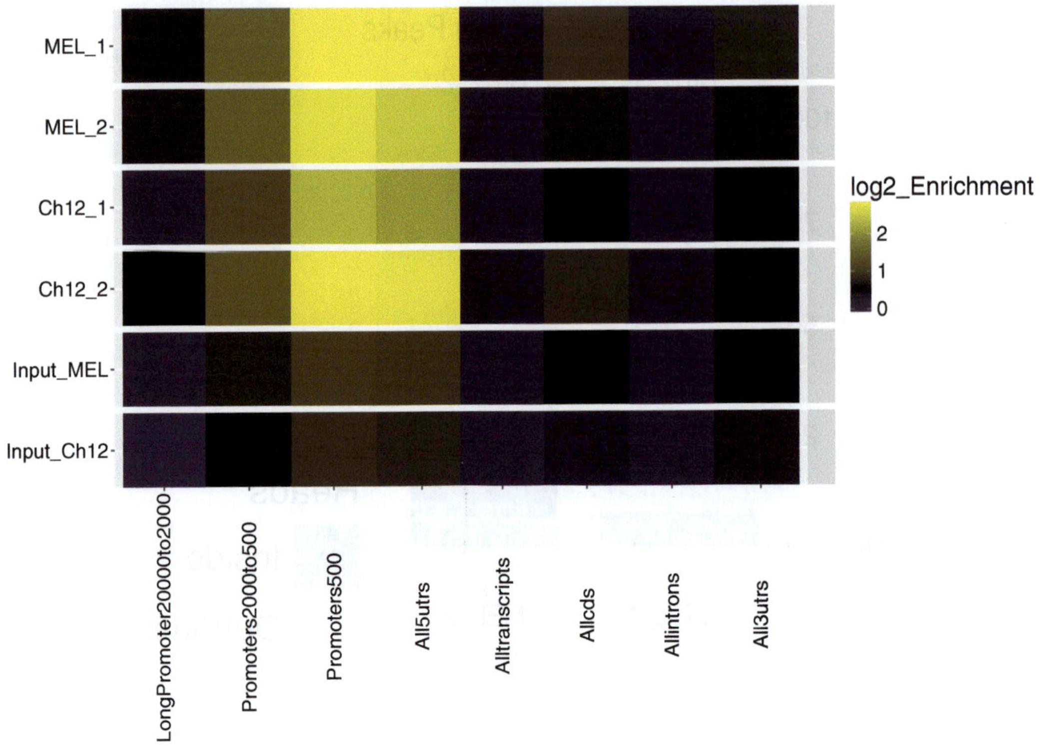

Fig. 6 Example ChIPC heatmap plot showing relative enrichment of genomic features

Bioconductor package DiffBind [21, 22]. Diffbind implements several normalization techniques and generates a variety of useful reports and plots.

First, we need to create a sample sheet with information about the localization of the BAM and BED files, as well as other metadata associated with the experiment.

```
#Create a samplesheet to set up the example cMyc experiment to analyze
samplesheet <- data.frame("sampleID"=
                    c("MEL_1","MEL_2","CH12_1","CH12_2"),
                    "Tissue"=c("MEL", "MEL","CH12","CH12"),
                    "Factor"=rep("cMyc",4),
                    "bamReads"=testSampleBAMs,
                    "bamControl"=testSampleControls,
                    "Peaks"=file.path("peaks",
                     paste0(namesForMACS2, "_peaks.narrowPeak"))
                    )
```

The function *dba.count* is used to calculate a binding matrix with scores based on read counts for every sample.

```
library(DiffBind)
dbsites <- dba(sampleSheet=samplesheet, peakCaller="narrow")
dbsites <- dba.count(dbsites)
```

Next, the *dba.contrast* function will tell DiffBind which samples fall into which groups.

```
dbsites <- dba.contrast(dbsites, minMembers=2)
```

The code below shows a short example code used to obtain the significantly differentially bound sites between Ch12 and MEL. This will run a DESeq2 [23] analysis by default.

```
dbsites <- dba.analyze(dbsites)
```

The final step is to retrieve the differentially bound sites with the function *dba.report.*

```
dbsites <- dba.report(dbsites)
```

DiffBind uses a negative binomial model to estimate the over-dispersion of reads between samples using statistical routines developed in an RNA-seq context (primarily the Bioconductor packages edgeR [24] and DESeq2 [23]). The main difference between the two methods is in the normalization of raw read counts (*see* **Note 6**). Similarly to DiffBind [21], the R package DBChIP [25] directly applies existing methods and software package developed for RNA-seq differential expression (DE) analysis. In general, a DE analysis works by computing normalization factors (or size factors) for individual samples that are used to re-scale the raw-counts and construct the model. However, this normalization methodology fails to consider that genome-wide distribution of read counts may have unequal mean and variance across samples (e.g., the different signal-to-noise ratios between samples). Other R packages, such as MAnorm [26] and ChIPcomp [27] try to address this issue. MAnorm focuses only on common peaks between samples as a reference to build the rescaling model for normalization. ChIPcomp performs quantitative comparison of multiple ChIP-seq datasets with considerations of data from the control experiment and differences in signal-to-noise ratios between experiments. Both methods assume that there is no global difference in the true biological signals within the common peaks across all datasets.

3.6 Visualization of ChIP-Seq Signal

3.6.1 Average Profiles Over Sites of Interest

The visualization of ChIP-seq signal over sites of interest is often performed to evaluate regions by their different signal profiles. In this example, we have generated sets of peaks which were found to be differentially bound between Ch12 and MEL and so here we will visualize the aggregated signal over peaks preferentially bound in either cell-line.

The soGGi [28] R Bioconductor package provided a set of tools to measure ChIP-seq signal as well as motif occurrence over absolute or length-normalized sets of genomic regions such as seen with meta-gene plots. To plot signal over differentially bound (DB) regions using the soGGi package, we must first update the GRanges object of DB regions with a column for soGGi to group by, here a column called "Class."

```
dbsites$Class <- ifelse(dbsites$Fold > 0, "Up_In_Ch12", "Up_In_MEL")
```

We use the *regionPlot* function in soGGi to plot the ChIP-seq signal over the DB regions. The *regionPlot* function accepts multiple input types for ChIP-seq signal and so here we provide the sorted, indexed BAM files for the ChIP-seq replicates, the DB sites, the fragment lengths for replicates from ChIPQC and specify the method as "point" to calculate the signal around the region in absolute base pairs.

```
library(soGGi)
register(BiocParallel::MulticoreParam(workers=numParallelJobs))
tt <- bplapply(1:length(allBams),function(x) {
      regionPlot(allBams[x],
             dbsites,
             method="point",
             FragmentLength=fragmentlength(chipqcList[[x]]))
  })
```

Having calculated the ChIP-seq signal of all samples over the differentially bound sites, we can visualize the metaprofile plot of mean signal over peaks using the *plotRegion* function from soGGi. Having updated the dbsites object to include their regulation within the relevant cell lines, we can use the *colourBy* and *groupBy* parameters to arrange the plots aesthetics by values within the metadata specified by the *summariseBy* parameters. The generated plot in Fig. 7 shows the mean ChIP-seq signal for all samples and inputs over two classes of c-Myc peaks (higher ChIP signal in MEL or Ch12). All samples show the presence of peak-shaped signal over the region centers with the expected differences in ChIP-seq signal for those peaks differentially bound between cell-lines. Input samples can be seen to have a minor increase in signal under their respective cell lines, illustrating the importance of the input subtraction used within DiffBind.

```
ttL <- do.call(c, tt)
plotRegion(ttL, colourBy="Sample",
         groupBy="Group", summariseBy="Class")
```

So far, we have plotted non-normalized ChIP-seq signal but soGGi provides functionality for normalization and arithmetic operations such as input subtraction. We may wish to normalize for differences in depth and ChIP efficiency and so here we apply a

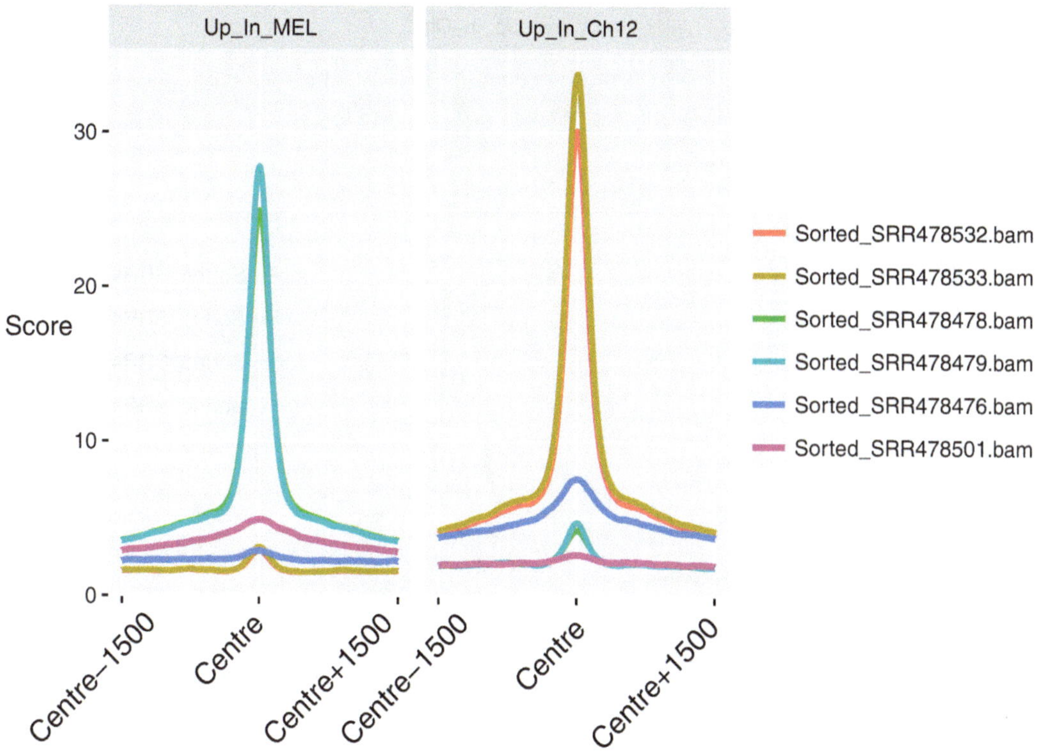

Fig. 7 Example soGGi average profile plot

quantile normalization and reproduce our previous plot with now normalized ChIP-seq signal over differential regions (Fig. 8).

```
sampleGroups <- c(tt[[1]], tt[[2]], tt[[3]], tt[[4]])
ttL_Quant <- normaliseQuantiles(sampleGroups)
plotRegion(ttL_Quant, colourBy="Sample",
           groupBy="Group", summariseBy="Class")
```

Following this, we can pool the replicate profiles from soGGi and produce a simplified normalized plot (Fig. 9).

```
Ch12_Profile <- ttL_Quant[[1]]+ttL_Quant[[2]]
metadata(Ch12_Profile)$names <- "Ch12_Samples"
Mel_Profile <- ttL_Quant[[3]]+ttL_Quant[[4]]
metadata(Mel_Profile)$names <- "MEL_Samples"
SummedProfiles <- c(Ch12_Profile,Mel_Profile)

plotRegion(SummedProfiles, colourBy="Sample",
           groupBy="Group", summariseBy="Class")
```

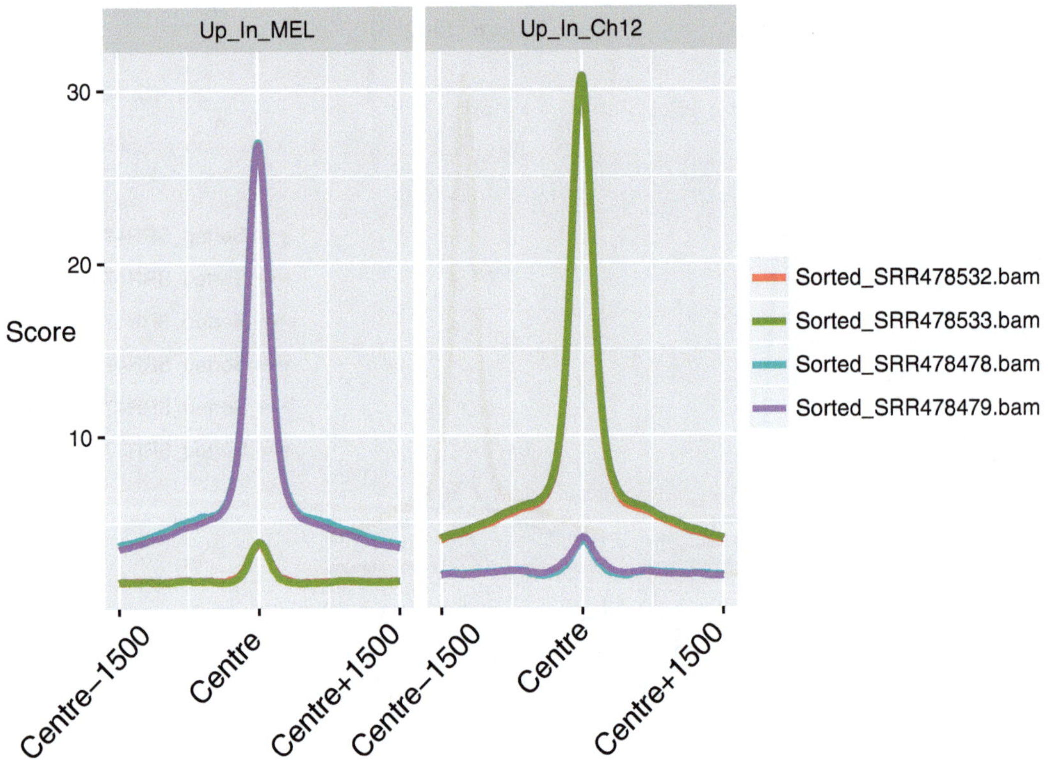

Fig. 8 Example soGGi average profile plot after quantile normalization

3.6.2 Data Visualization Through Coverage Tracks

Having aligned the data, called peaks as well as identified those peaks showing differential Myc signal between MEL and Ch12 cell lines, we will wish to visualize our results. A common approach to visualizing genomics data is to make use of the interactive, linear representation of genomic information implemented in genome browsers.

Many genome browsers have been created to visualize a range of genomics data types including the server hosted UCSC browser and the locally hosted, Java-based Integrative Genome Viewer (IGV) browser. Here, we will make use of the locally hosted IGV to visualize our ChIP-seq signal and enriched/differentially enriched regions.

Although IGV allows for the visualization of BAM files, displaying their read pileup, quality and strand, for HTS it is often more efficient to calculate the signal coverage upfront and store this information in a bigWig file. This would also allow us to extend reads to fragment length calculated by ChIPQC and hence construct a bigWig with a more accurate representation of predicted ChIP signal than seen from BAM files.

To create the read-extended bigWigs, we make use of the *readGAlignments* function to collect reads and extend them to the fragment length using the *resize* function. Having resized the

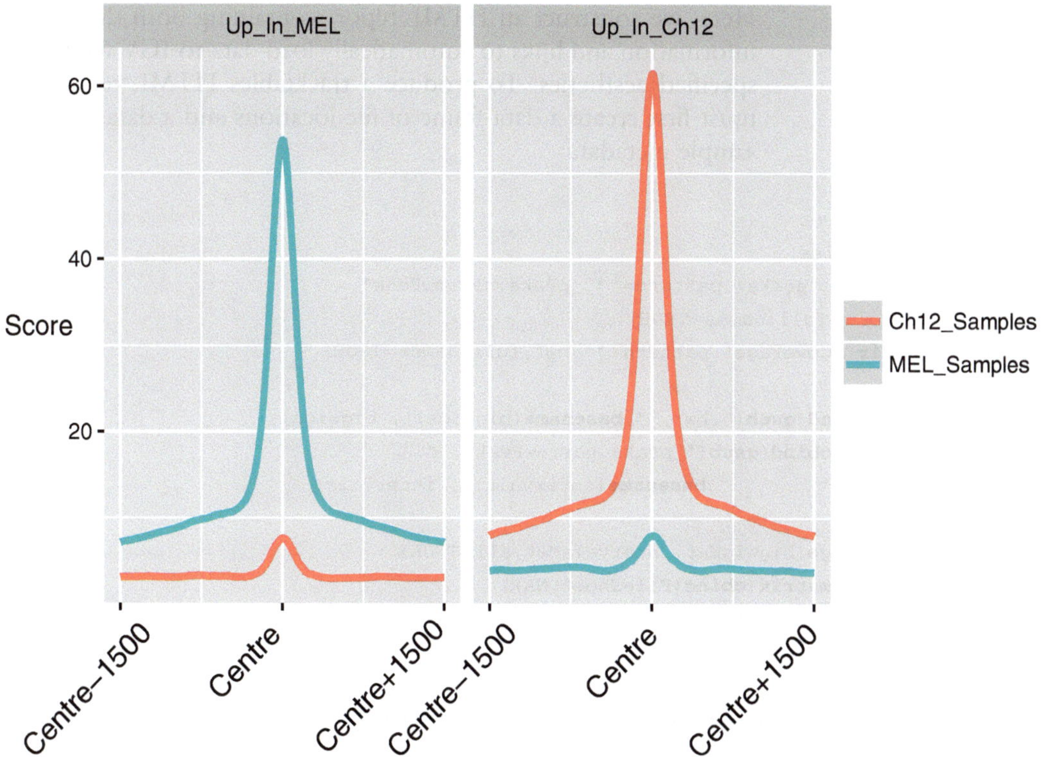

Fig. 9 Simplified soGGi average profile plot after quantile normalization

reads to predicted fragments, we produce the signal coverage using the *coverage* function and export to bigWig using the *export.bw* function in the rtracklayer package.

```
library(GenomicAlignments)
library(rtracklayer)
dir.create("Coverage",showWarnings=FALSE)
bigWigs <- basename(gsub("\cr.bam","\cr.bw", allBams))

lapply(1:length(allBams),
         function(x)
           export.bw(coverage(
              resize(GRanges(readGAlignments(allBams[x])),
                 fragmentlength(chipqcList)[x])),
                        con=file.path("Coverage",bigWigs[x])))
```

The much required organization of experimental metadata within genome browsers has grown in complexity alongside the increase in experimental size and data diversity. IGV allows the user to specify sample metadata and low-level configuration of options using XML and more complex custom file types.

The tracktables [29] package allows for the construction of IGV-linked HTML reports with a set of user-friendly R commands.

Here, we construct an HTML report containing both the sample information and links to automatically load data to IGV with user-specified aesthetics. To produce a tracktables HTML report, we must first create a data-frame of file locations and a data-frame of sample metadata.

```
#Create a FileSheet
library(tracktables)
intervals <- dir("peaks",pattern="*._peaks.narrowPeak",
                 full.names=TRUE)
bigwigs <- dir("Coverage",pattern="*.bw",full.names=TRUE)

bigWigMat <- cbind(gsub(".bw","",basename(bigwigs)), bigwigs)
intervalsMat <- cbind(gsub("_peaks.narrowPeak","",
                           basename(intervals)), intervals)

FileSheet <- merge(bigWigMat,intervalsMat,all=TRUE)
FileSheet <- as.matrix(cbind(FileSheet,NA))
colnames(FileSheet) <- c("SampleName","bigwig","interval","bam")

#Create a SampleSheet
SampleSheet <- cbind(as.vector(FileSheet[,"SampleName"]),
                     c("MEL","Ch12","Ch12","Ch12","MEL","MEL"),
                     c("Input","Myc","Myc","Input","Myc","Myc"))
colnames(SampleSheet) <- c("SampleName","Cell-line","Antibody")
```

Having created these, we use the *maketracktable* function with the user-defined *igvParam* as well as specify coloring by antibody by setting the *colourBy* parameter to *Antibody*. Finally, we can create an HTML of differentially bound regions using the *makebedtable* function.

```
igvDisplayParams <- igvParam(bigwig.autoScale="true")

HTMLreport <- maketracktable(FileSheet,SampleSheet,
                    "Tracktable_For_ChIPseq.html",getwd(),"mm9",
                     colourBy="Antibody",igvParam=igvDisplayParams)

makebedtable(dbsites,"HTMLofDBSites.html",getwd())
```

3.7 Functional Downstream Analysis

3.7.1 Peak Annotation: Determination of Target Genes

A common goal in ChIP-seq is to associate putative transcription factor (TF) sites to their target genes. Many methods for associating TF binding sites to genes have been implemented including the association of TFs to genes within a set distance, to the nearest genes [30], to genes within Topologically Associated Domains or through using further methods measuring long-range DNA interactions (5C, Hi-C) [31].

In this example, we will use the ChIPseeker [32] Bioconductor package to associate peaks to their nearest gene's transcriptional start site and the position within features of a genes. Here, we annotate the differentially bound sites using the *annotatePeak* function specifying the distance around a gene's start to define as TSS.

```
library(TxDb.Mmusculus.UCSC.mm10.knownGene)
library(org.Mm.eg.db)
library(GenomeInfoDb)
library(ChIPseeker)

Up_In_Ch12 <- dbsites[dbsites$Class == "Up_In_Ch12"]
Up_In_MEL  <- dbsites[dbsites$Class == "Up_In_MEL"]

Up_In_Ch12_Anno <- annotatePeak(
                            Up_In_Ch12, tssRegion=c(-1000, 1000),
                            TxDb=TxDb.Mmusculus.UCSC.mm10.knownGene,
                            annoDb="org.Mm.eg.db")

Up_In_MEL_Anno <- annotatePeak(
                            Up_In_MEL,
                            tssRegion =c(-1000, 1000),
                            TxDb=TxDb.Mmusculus.UCSC.mm10.knownGene,
                            annoDb="org.Mm.eg.db")
```

The ChIPseeker package offers functionality to visualize annotation of a peak set. To retrieve the peak sets with genes annotated we can apply the *as.GRanges* function to the *csanno* object returned from *annotatePeak.*

```
Up_In_Ch12WithGenes <- as.GRanges(Up_In_Ch12_Anno)
Up_In_MelWithGenes <- as.GRanges(Up_In_MEL_Anno)
```

To further investigate the annotation of peak sets, we apply the *plotAnnoBar* function to the list of *csanno* objects. We may also visualize the distance of peaks around transcription factor binding sites using *plotDistToTSS* function.

From these two plots, in Figs. 10 and 11, it is apparent that the peaks sets have very different patterns of binding. The peaks with higher signal in Ch12 were found to have a higher proportion of peaks in genes' promoters.

```
DBpeaks <- list(Up_In_Ch12_Anno, Up_In_MEL_Anno)
names(DBpeaks) <- c("Up_In_Ch12", "Up_In_MEL")
plotAnnoBar(DBpeaks)
plotDistToTSS(DBpeaks)
```

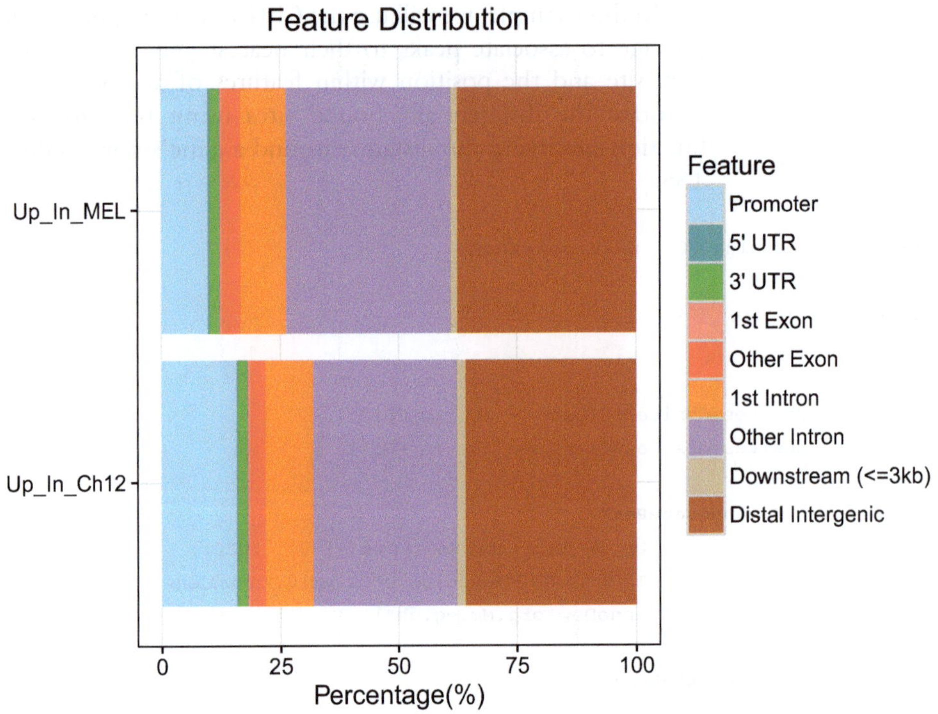

Fig. 10 Example ChIPseeker peak annotation plot

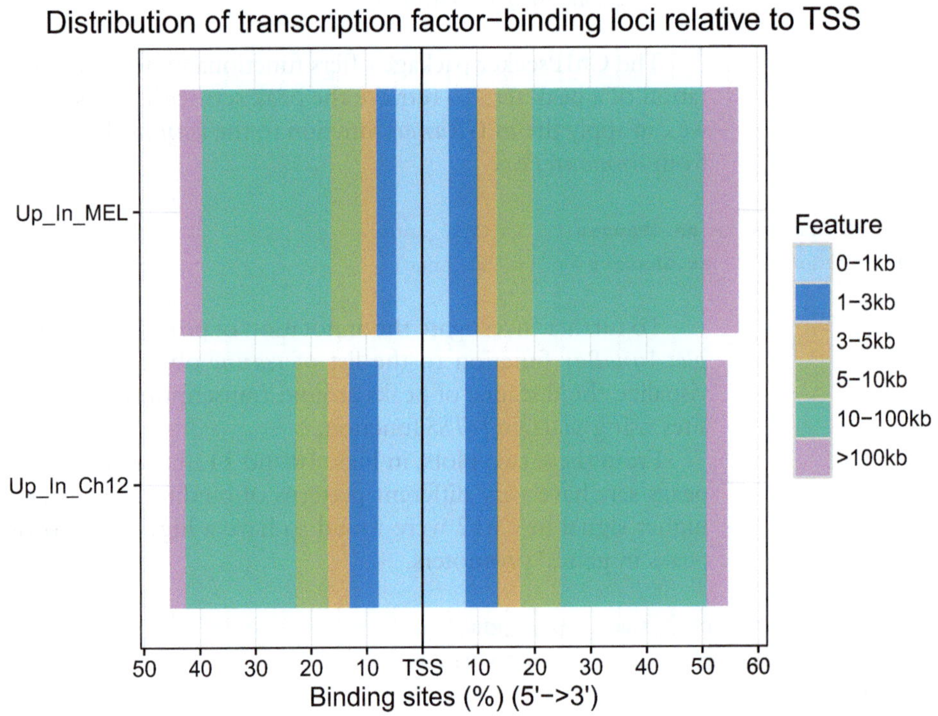

Fig. 11 Example ChIPseeker plot showing the distribution of TF-binding loci relative to TSS

3.7.2 Enrichment Within Functional Categories

Following annotation of transcription factor binding sites to target genes, it is possible to test for an enrichment of these gene targets within functional categories as commonly performed for RNA-seq analysis. Repositories such as those of the Gene Ontology, Reactome Pathway Database, and Broads MSigDB contain sets of genes associated with a common process, pathway, or other biological annotation. R Packages such as ChIP-Enrich [33] use predefined gene sets to perform gene-set enrichment analysis after assigning peaks to genes. Although this type of analysis can be applied to promoter-centric gene annotation of ChIP-seq data (from which the gene is predicted to be regulated), such statistics are not applicable to the annotation of peaks to nearest genes or variable regulatory regions. To address this, the use of statistics accounting for genes' regulatory regions' sizes has been employed in the testing of gene sets and functional categories for ChIP-seq data [34].

In this example, we use the rGREAT package [34] to test the enrichment for gene targets of GO Biological Processes. We use the *submitGreatJob* function to submit our peaks to the GREAT server specifying the genome in the *species* parameter.

```
library(rGREAT)
great_Job_Ch12 <- submitGreatJob(Up_In_Ch12, species="mm10")
great_Job_Mel <- submitGreatJob(Up_In_MEL, species="mm10")
```

Having submitted the peaks, we browse the list of available functional categories with *availableCategories* and here selecting the GO set of results using the *getEnrichmentTables* function.

```
availableCategories(great_Job_Ch12)
availableCategories(great_Job_Mel)
great_ResultTable_Ch12 <- getEnrichmentTables(great_Job_Ch12,
                                              category="GO")
great_ResultTable_Mel <-getEnrichmentTables(great_Job_Mel,
                                            category="GO")
```

Finally, we can select the results from the Biological Processes branch from the results list. In this example, we can see that both sets of peaks are enriched for gene targets within the metabolic pathways as expected by Myc's known functions. Interestingly, the MEL cell-line peaks show a significantly stronger enrichment for the "immune system process" category (Table 1) than those in Ch12 (Table 2) highlighting an important difference of these two B-Cell lymphoma derived cell-lines.

Table 1
Gene ontology results for MEL cell-line peaks (top six according to hypergeometric *p*-value)

GO ID	Name	Hypergeometric *p*-value
GO:0002376	Immune system process	2.34e−27
GO:0009987	Cellular process	2.45e−22
GO:0048583	Regulation of response to stimulus	6.68e−63
GO:0050789	Regulation of biological process	1.84e−07
GO:0050794	Regulation of cellular process	7.05e−08
GO:0019222	Regulation of metabolic process	1.76e−43

Table 2
Gene ontology results for CH12 cell-line peaks (top six according to hypergeometric *p*-value)

GO ID	Name	Hypergeometric *p*-value
GO:0044237	Cellular metabolic process	2.94e−106
GO:0008152	metabolic process	1.14e−82
GO:0044238	Primary metabolic process	3.90e−80
GO:0071704	Organic substance metabolic process	3.11e−85
GO:0009987	Cellular process	8.54e−39
GO:0006807	Nitrogen compound metabolic process	6.09e−54

```
msigdb_great_ResultTable_Ch12 <-
                       great_ResultTable_Ch12[["GO Biological Process"]]
msigdb_great_ResultTable_Mel <-
                       great_ResultTable_Mel[["GO Biological Process"]]

head(msigdb_great_ResultTable_Mel)
head(msigdb_great_ResultTable_Ch12)
```

3.7.3 Motif Finding

A common step in a ChIP-seq analysis is to identify binding motifs associated with the peak set [35, 36].

To identify binding motifs, we will first need to extract FASTA files around the centers of peaks. To get the genomic coordinates of windows of 150 bp around the peak centers, we use the function *resize* from GenomicRanges [37] package and the list of peak sets in the GRanges format.

```
library(GenomicRanges)
peaks <- file.path("peaks",
         c("CH12_cMyc1"="Sorted_SRR478478_peaks.narrowPeak",
           "CH12_cMyc2"="Sorted_SRR478479_peaks.narrowPeak",
           "MEL_cMyc1"="Sorted_SRR478532_peaks.narrowPeak",
           "MEL_cMyc2"="Sorted_SRR478533_peaks.narrowPeak"))

peaksGR <- lapply(peaks, function(x) {
    x <- read.delim(x, header=FALSE)
    colnames(x) <- c("chrom", "chromStart", "chromEnd", "name",
                     "score", "strand", "fold.enrichment",
                     "log10.pvalue","log10.qvalue", "peak")
    peaksGR <- GRanges(x$chrom, IRanges(x$chromStart+1, x$chromEnd))
    peaksGR
})

peakCentres <- lapply(peaksGR, resize, fix="center", width=150)
```

The function *getAllPeakSequence* from ChIPpeakAnno [30] package is then used to retrieve the genomic sequences at these coordinates. For this, the ChIPpeakAnno requires the annotation contained in the BSgenome.Mmusculus.UCSC.mm10 [38] package which will also need to be loaded.

```
library(BSgenome.Mmusculus.UCSC.mm10)
library(ChIPpeakAnno)

peakSeq <- getAllPeakSequence(peakCentres[[1]],
                              upstream=0,
                              downstream=0,
                              genome=Mmusculus)
sequences <- peakSeq$sequence
names(sequences) <- paste(peakSeq$space,
                          start(peakSeq),
                          end(peakSeq), sep="_")
```

Finally, we use the *DNAStringSet* function of the Biostrings package to retrieve the DNA sequences as a *XStringSet* object, and *seqXstring* function to write to a file.

```
seqXstring <- DNAStringSet(sequences)
writeXStringSet(seqXstring, filepath="sequences.fasta")
```

The generated FASTA file can be uploaded directly in the Meme-ChIP suite (http://meme-suite.org/tools/meme-chip) [35] website to identify the enriched TF motifs within the peak set. First, go to the Meme-ChIP website, and choose "upload sequences." Choose the "sequences.fasta" file from the results folder. Then click "start search." Once the results are ready, they will be available on the "Recent Jobs" menu on the left.

3.7.4 Motif-Based Evaluations of Peak Sets

Motif analysis using Meme-ChIP suite identified the expected Myc motif within our Myc peaks. We can visualize the strength at the ChIP-seq signal at the c-Myc known binding motif site as a measure of the accuracy of peak calling results (*see* **Note 7**).

The first step is to identify Myc motifs within our peaks. For that we will need to provide details of nucleotide frequency at positions within the motif, often as a Point Weight Matrix (PWM). We can retrieve PWMs from several sources including Transfac [39] and Jasper [40]. Here, we will use the MotifDb [41] package to obtain the Japsar-2014 Myc motif as a PWM.

```
library(MotifDb)
PotentialMyc <- as.list(subset (MotifDb,
                          tolower (geneSymbol)=="myc"))
```

Having found the motif of interest, we can then visualize it as a sequence logo using the SeqLogo [42] package. Sequence logos, as seen below in Fig. 12 for c-Myc, visualize the information content for each nucleotide at every position within the motif.

```
library(seqLogo)
seqLogo(PotentialMyc[[10]])
```

With the appropriate PWM, we can scan the genome of interest for the occurrence of motifs. Here, we scan the genome sequence contained within BSgenome.Mmusculus.UCSC.mm10 Bioconductor package using the *pwmToCoverage* function in soGGi.

```
library(BSgenome.Mmusculus.UCSC.mm10)
resMyc <- pwmToCoverage(PotentialMyc[[10]],
                        BSgenome.Mmusculus.UCSC.mm10,
                        min="70%", removeRand=TRUE,
                        chrsOfInterest=paste0("chr",c(1:19,"X","Y")))
```

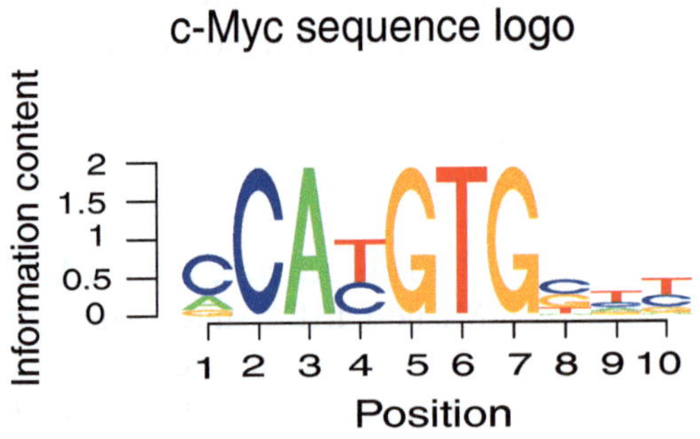

Fig. 12 Sequence logo for c-Myc

Since motifs for the ChIP of interest are largely expected to be at the summit of ChIP signal within a peak, we will first identify summits for our differentially bound regions. Identifying summits for consensus peaks such as those from DiffBind will require information of the signal from multiple samples. We can use the *findconsensusRegions* function from soGGi to identify the summit for all peaks using the mean of all samples' summits, weighted by the respective sample's normalized summit height. Here we use the *findconsensusRegions* function with our differential bound sites, specifying the BAM files for ChIP-seq signal and the parameter *summit* as *weightedmean*.

```
testSampleFragmentLengths <- fragmentlength(chipqcList)[1:4]
names(testSampleBAMs) <- names(testSampleFragmentLengths)

dbsitesCentred <- findconsensusRegions(dbsites,testSampleBAMs,
            method="majority", summit="weightedmean",
            resizepeak="asw", overlap="any",
            fragmentLength=testSampleFragmentLengths,
            NonPrimaryPeaks=list(withinsample="drop",
            betweensample="weightedmean"),
            calculateSinglePeakSummit=TRUE, regionAsIs=TRUE)
```

Having re-summited our differentially bound sites, we can use the *rlelist* of motif occurrence generated by the *pwmToCoverage* function to produce a metaprofile of motif frequency. Here, we use the *regionPlot* function again but with the parameter of *format* set to *rlelist*. To plot the motif frequency metaprofile, we can again use the *plotRegion* function. In the resulting profile, in Fig. 13, we can see the expected sharp peak of Myc frequency at the peak summits' indicating a successful ChIP-seq experiment.

```
ttMotifCentred <- regionPlot(resMyc, dbsitesCentred, style="point",
                              format="rlelist")
plotRegion(ttMotifCentred)
```

4 Notes

1. For quality control, the inclusion of "best hit" for multiple mapping reads is recommended to capture the blacklist artifact effects.
2. It is recommended that reads mapping to multiple locations are removed prior to peak calling with MACS2, to prevent the detection of false-positive peaks.

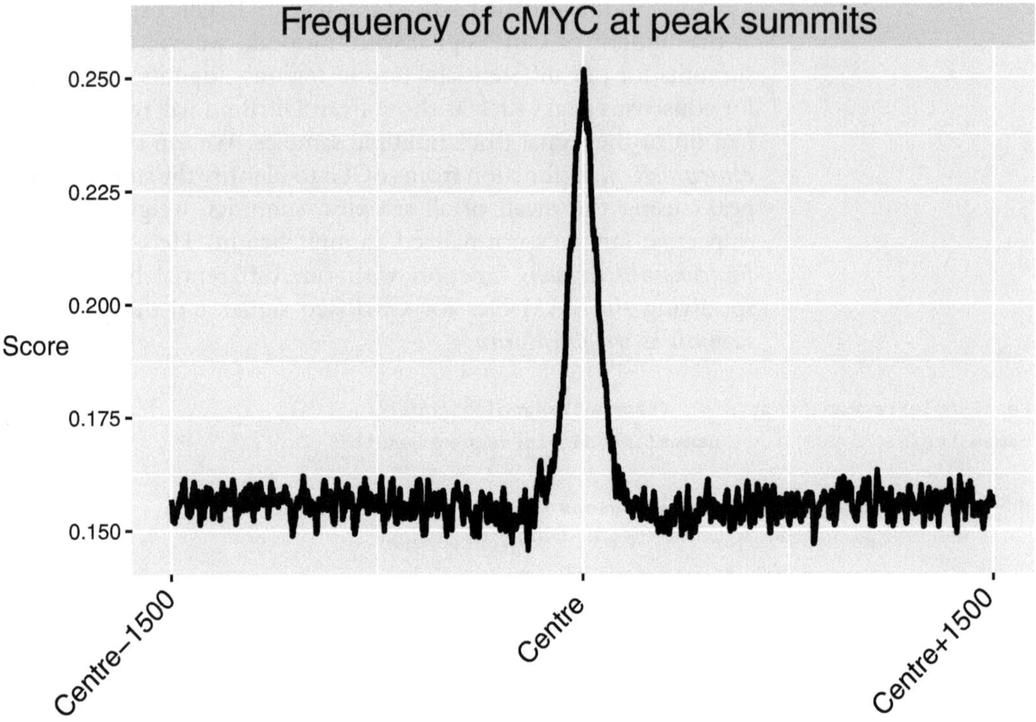

Fig. 13 Average profile plot at c-Myc motifs

3. MACS2 is a widely used peak caller. However, the appropriate method largely depends on the type of signal of the targeted proteins. MACS2 has parameter settings for both sharp and broader modes of ChIP enrichment signals. There are many other peak callers that have been specifically developed for broader (e.g., H3K9me3) or mixed (e.g., RNA-PolII) peak regions (e.g., ZINBA [43], BroadPeak [44], MUSIC [45]), however such detection is still challenging.
4. In a differential binding analysis, biological replicates are essential to characterize variability between samples.
5. In studies involving the comparison of multiple samples, it is important to look out for batch effects. Batch effects may be conflated with treatment effects of interest [46].
6. Choosing an appropriate normalization scheme (e.g., DESeq2 vs edgeR) can be challenging. It is often good practice to assess if results are consistent using different normalization methods.
7. In a Motif-based evaluation analysis, it is important to note that even proteins with a known canonical motif can have other binding sites that do not involve that specific motif sequence [47].

References

1. Landt SG, Marinov GK, Kundaje A et al (2012) ChIP-seq guidelines and practices of the ENCODE and modENCODE consortia. Genome Res 22(9):1813–1831
2. Meyer CA, Liu XS (2014) Identifying and mitigating bias in next-generation sequencing methods for chromatin biology. Nat Rev Genet 15(11):709–721
3. Consortium EP (2012) An integrated encyclopedia of dNA elements in the human genome. Nature 489(7414):57–74
4. .NCBI fTP website for mm10 genome (2016) ftp://ftp.ncbi.nih.gov/genomes/genbank/vertebrate_mammalian/Mus_musculus/all_assembly_versions/GCA_000001635.7_GRCm38.p5/GCA_000001635.7_GRCm38.p5_assembly_structure/Primary_Assembly/assembled_chromosomes/FASTA. Accessed 8 Mar 2017
5. Zhu Y, Stephens RM, Meltzer PS et al (2013) SRAdb: query and use public next-generation sequencing data from within R. BMC Bioinformatics 14:19
6. Pagès H, Aboyoun P, Gentleman R, DebRoy S (2016) Biostrings: string objects representing biological sequences, and matching algorithms. R package version. 2.42.1
7. Liao Y, Smyth GK, Shi W (2013) The subread aligner: fast, accurate and scalable read mapping by seed-and-vote. Nucleic Acids Res 41:e108
8. Lawrence M, Gentleman R, Carey V (2009) Rtracklayer: an R package for interfacing with genome browsers. Bioinformatics 25 (14):1841–1842
9. Brown G (2016) GreyListChIP: grey lists – mask artefact regions based on ChIP inputs. R package version 1.4.1
10. Langmead B, Trapnell C, Pop M et al (2009) Ultrafast and memory-efficient alignment of short DNA sequences to the human genome. Genome Biol 10(3):R25
11. Gaidatzis D, Lerch A, Hahne F et al (2015) QuasR: quantification and annotation of short reads in R. Bioinformatics 31(7):1130–1132
12. Morgan M, Pagès H, Obenchain V, et al (2016) Rsamtools: binary alignment (BAM), FASTA, variant call (BCF), and Tabix file import. R package version 1.26.1
13. Park PJ (2009) ChIP–seq: advantages and challenges of a maturing technology. Nat Rev Genet 10(10):669–680
14. Zhang Y, Liu T, Meyer CA et al (2008) Model-based analysis of ChIP-seq (MACS). Genome Biol 9(9):R137
15. Kharchenko PV, Tolstorukov MY, Park PJ (2008) Design and analysis of ChIP-seq experiments for DNA-binding proteins. Nat Biotechnol 26(12):1351–1359
16. Hower V, Evans SN, Pachter L (2011) Shape-based peak identification for ChIP-seq. BMC Bioinformatics 12:1
17. Kornacker K, Rye MB, Haandstad T et al (2012) The triform algorithm: improved sensitivity and specificity in ChIP-seq peak finding. BMC Bioinformatics 13:1
18. Carroll TS, Liang Z, Salama R et al (2014) Impact of artifact removal on chIP quality metrics in ChIP-seq and ChIP-exo data. Front Genet 5:75
19. Planet E, Attolini CS-O, Reina O et al (2012) htSeqTools: high-throughput sequencing quality control, processing and visualization in R. Bioinformatics 28(4):589–590
20. Liu T, Ortiz JA, Taing L et al (2011) Cistrome: an integrative platform for transcriptional regulation studies. Genome Biol 12(8):R183
21. Stark R, Brown G (2011) DiffBind: Differential binding analysis of ChIP-seq peak data. R package version 2.2.8
22. Ross-Innes CS, Stark R, Teschendorff AE et al (2012) Differential oestrogen receptor binding is associated with clinical outcome in breast cancer. Nature 481(7381):389–393
23. Love MI, Huber W, Anders S (2014) Moderated estimation of fold change and dispersion for RNA-seq data with DESeq2. Genome Biol 15(12):550
24. Robinson MD, McCarthy DJ, Smyth GK (2010) EdgeR: a bioconductor package for differential expression analysis of digital gene expression data. Bioinformatics 26 (1):139–140
25. Liang K (2012) DBChIP: Differential Binding of Transcription Factor with ChIP-Seq. R package version 1.16.0
26. Shao Z, Zhang Y, Yuan G-C et al (2012) MAnorm: a robust model for quantitative comparison of ChIP-seq data sets. Genome Biol 13(3): R16
27. Chen L, Wang C, Qin ZS et al (2015) A novel statistical method for quantitative comparison of multiple ChIP-seq datasets. Bioinformatics 31(12):1889–1896
28. Dharmalingam G, Carroll T (2015) SoGGi: visualise ChIP-Seq, MNase-Seq and motif occurrence as aggregate plots summarised over grouped genomic intervals. R package version 1.4.4

29. Carroll T, Khadayate S, Pajon A, et al (2014) Tracktables: build IGV tracks and HTML reports. R package version 1.8.1
30. Zhu LJ, Gazin C, Lawson ND et al (2010) ChIPpeakAnno: a bioconductor package to annotate ChIP-seq and ChIP-chip data. BMC Bioinformatics 11:1
31. Heinz S, Benner C, Spann N et al (2010) Simple combinations of lineage-determining transcription factors prime cis-regulatory elements required for macrophage and b cell identities. Mol Cell 38(4):576–589
32. Yu G, Wang L-G, He Q-Y (2015) ChIPseeker: an R/Bioconductor package for ChIP peak annotation, comparison and visualization. Bioinformatics 31(14):2382–2383
33. Welch RP, Lee C, Imbriano PM et al (2014) ChIP-enrich: gene set enrichment testing for ChIP-seq data. Nucleic Acids Res 42(13):e105
34. McLean CY, Bristor D, Hiller M et al (2010) GREAT improves functional interpretation of cis-regulatory regions. Nat Biotechnol 28 (5):495–501
35. Machanick P, Bailey TL (2011) MEME-ChIP: motif analysis of large DNA datasets. Bioinformatics 27(12):1696–1697
36. Thomas-Chollier M, Herrmann C, Defrance M et al (2012) RSAT peak-motifs: motif analysis in full-size ChIP-seq datasets. Nucleic Acids Res 40(4):e31
37. Lawrence M, Huber W, Pagés H et al (2013) Software for computing and annotating genomic ranges. PLoS Comput Biol 9(8):e1003118
38. Team TBD (2014) BSgenome.Mmusculus. UCSC.mm10: full genome sequences for mus musculus UCSC version mm10. R package version 1.4.0
39. Matys V, Kel-Margoulis OV, Fricke E et al (2006) TRANSFAC and its module tRANSCompel: transcriptional gene regulation in eukaryotes. Nucleic Acids Res 34(Database issue):D108–D110
40. Mathelier A, Zhao X, Zhang AW et al (2013) JASPAR 2014: an extensively expanded and updated open-access database of transcription factor binding profiles. Nucleic Acids Res 42 (Database issue):D142–D147
41. Shannon P (2016) MotifDb: an annotated collection of protein-DNA binding sequence motifs. R package version 1.14.0.
42. Bembom O (2016) SeqLogo: sequence Logos for DNA sequence alignments. R package version 1.40.0
43. Rashid NU, Giresi PG, Ibrahim JG et al (2011) ZINBA integrates local covariates with DNA-seq data to identify broad and narrow regions of enrichment, even within amplified genomic regions. Genome Biol 12(7):R67
44. Wang J, Lunyak VV, Jordan IK (2013) BroadPeak: a novel algorithm for identifying broad peaks in diffuse ChIP-seq datasets. Bioinformatics 29(4):492–493
45. Harmanci A, Rozowsky J, Gerstein M (2104) MUSIC: identification of enriched regions in ChIP-seq experiments using a mappability-corrected multiscale signal processing framework. Genome Biol 15(10):474
46. Leek JT, Scharpf RB, Bravo HC et al (2010) Tackling the widespread and critical impact of batch effects in high-throughput data. Nat Rev Genet 11:733–739
47. Nakato R, Shirahige K (2016) Recent advances in ChIP-seq analysis: from quality management to whole-genome annotation. Brief Bioinform 8(2):279–290

Chapter 18

Spike-In Normalization of ChIP Data Using DNA–DIG–Antibody Complex

Andrea B. Eberle

Abstract

Chromatin immunoprecipitation (ChIP) is a widely used method to determine the occupancy of specific proteins within the genome, helping to unravel the function and activity of specific genomic regions. In ChIP experiments, normalization of the obtained data by a suitable internal reference is crucial. However, particularly when comparing differently treated samples, such a reference is difficult to identify. Here, a simple method to improve the accuracy and reliability of ChIP experiments by the help of an external reference is described. An artificial molecule, composed of a well-defined digoxigenin (DIG) labeled DNA fragment in complex with an anti-DIG antibody, is synthesized and added to each chromatin sample before immunoprecipitation. During the ChIP procedure, the DNA–DIG–antibody complex undergoes the same treatments as the chromatin and is therefore purified and quantified together with the chromatin of interest. This external reference compensates for variability during the ChIP routine and improves the similarity between replicates, thereby emphasizing the biological differences between samples.

Key words Chromatin immunoprecipitation (ChIP), Spike-in, Digoxigenin, DIG-11-dUTP, DNA–-DIG–antibody complex, Normalization, External reference

1 Introduction

Chromatin immunoprecipitation (ChIP) is a widely used method in molecular cell biology to identify the occupancy of DNA-bound proteins (such as histones, polymerases, and transcription factors) within the genome thereby pointing to the state and the function of DNA regions [1]. Next to identifying protein–DNA interaction directly, this method is also valuable to determine indirect interactions of proteins with the genome, whereby this indirect interaction can be mediated by other proteins or RNA (e.g., transcriptional co-activators, histone deacetylases or splicing factors, heterogeneous nuclear ribonucleoproteins (hnRNPs)) [2, 3]. The analysis of DNA sequences bound by the protein of interest is achieved by high-throughput sequencing for the whole genome (ChIP-seq) or by

Neus Visa and Antonio Jordán-Pla (eds.), *Chromatin Immunoprecipitation: Methods and Protocols*,
Methods in Molecular Biology, vol. 1689, DOI 10.1007/978-1-4939-7380-4_18, © Springer Science+Business Media LLC 2018

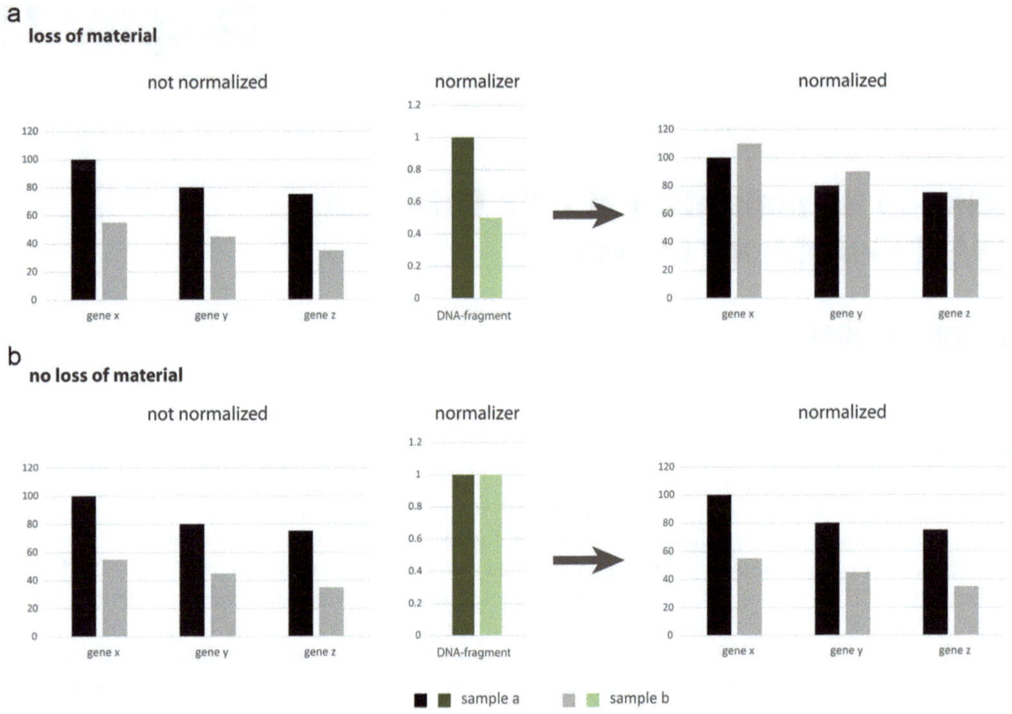

Fig. 1 Analysis of ChIP experiments to illustrate the importance of normalization. Here, a devised ChIP experiment with two independent samples (sample a and b) is shown. (**a**) Three different DNA regions (gene x, y, and z) are quantified in the immunoprecipitated samples by qPCR for both samples and shown relative to the input sample (*left panel*). The DNA–DIG–antibody complex was added as external reference to the chromatin extracts before immunoprecipitation and co-purified with the DNA of interest. The DNA fragment is analyzed by qPCR in sample a and b (in *green*, *middle panel*). In this scenario the DNA fragment is reduced half in sample b compared to sample a, indicating sample loss during chromatin isolation or DNA purification in sample b. When normalizing the data with the help of the external reference, no difference between sample a and b for gene x, y, and z can be determined (*right panel*). In (**b**) same experiment as in (**a**) is shown, but in this scenario no material was lost during the ChIP experiment. Therefore, the levels of the DNA fragment measured by qPCR are the same in sample a and b (*middle panel*). Using this reference for normalization, a clear difference between sample a and b is detected for gene x, y, and z

real-time PCR for specific regions of the genome (quantitative PCR, qPCR).

The ChIP method includes several critical aspects such as cross-linking of cells [4], specificity/suitability of antibodies [5, 6], and normalization issues that have to be addressed. Normalization is crucial to reduce experimentally induced variation between samples and thereby increases the reliability of ChIP data (Fig. 1). The use of an internal reference, in other words adjusting the data to an endogenous DNA region that is expected to be identical in all samples, is the favored normalization strategy, as this "normalizer" is already present in the sample from the beginning and thus treated exactly the same way as the chromatin of interest. In many experiments, however, the occupancy of a specific protein within the genome is compared in diverse cell types or different conditions (e.g., depletion/overexpression of proteins, treatments with drugs). To compensate for variations between these different samples, the use of an internal reference is often precluded since a DNA sequence that is not affected by the experimental routine cannot be determined and/or global effects are expected. Here, an external reference that does not interfere with the ChIP experiment helps to assess biological differences between samples by improving the statistical resolution of the data. The ideal external normalizer should resemble the biological samples in such a way that it can be purified and quantified by the same methods as the chromatin. An interesting approach is to spike the chromatin with a low amount of chromatin from a different species before immunoprecipitation and use the foreign chromatin as reference [7]. This so-called spike adjustment procedure is however restricted to proteins of interest that share conserved epitopes in both species and are therefore recognized by the same antibody.

Here, the protocol for the synthesis and application of an artificial molecule that can be used as external reference in ChIP experiments is described [8]. For this purpose, a DNA sequence of bacterial origin lacking homologies in eukaryotes is labeled with digoxigenin (DIG) and incubated with a stoichiometric amount of an anti-DIG antibody. The formed complex, composed of the antibody and the DIG-labeled DNA, is crosslinked with formaldehyde and purified. At the beginning of the ChIP experiment, a fixed amount of the purified DNA–DIG–antibody complex is added to the harvested and sheared chromatin before the incubation with the antibody against the protein of interest. The addition of this artificial reference does not disturb the subsequent ChIP procedure, i.e., isolation of chromatin bound to antibodies with the help of protein A and/or G coupled to magnetic or sepharose beads, digestion of proteins and RNA, reversal of the crosslinking, and purification of the DNA (Fig. 2). Finally, all purified DNA sequences, including the DNA fragment of the spiked DNA–DIG–antibody complex, are quantified in both the starting material (input) and in the immunoprecipitated sample by qPCR or sequencing. Importantly, the DNA–DIG–antibody complex serves as normalizer to compensate for variations in different samples introduced during chromatin isolation, reversal of the crosslinking or DNA purification (Fig. 1). The normalization of the ChIP signals thus minimizes variability of technical origin between the individual samples and increases robustness and accuracy of the data. It is noteworthy however that variation referring to the quality of the chromatin cannot be corrected by this method.

The DNA–DIG external reference is universal and can be applied in any ChIP experiment containing chromatin from animal cells. The chosen DNA sequence is a part of the quinol *bo3* oxidase of *Escherichia coli* (*E. coli*) that is absent in eukaryotic genomes. The

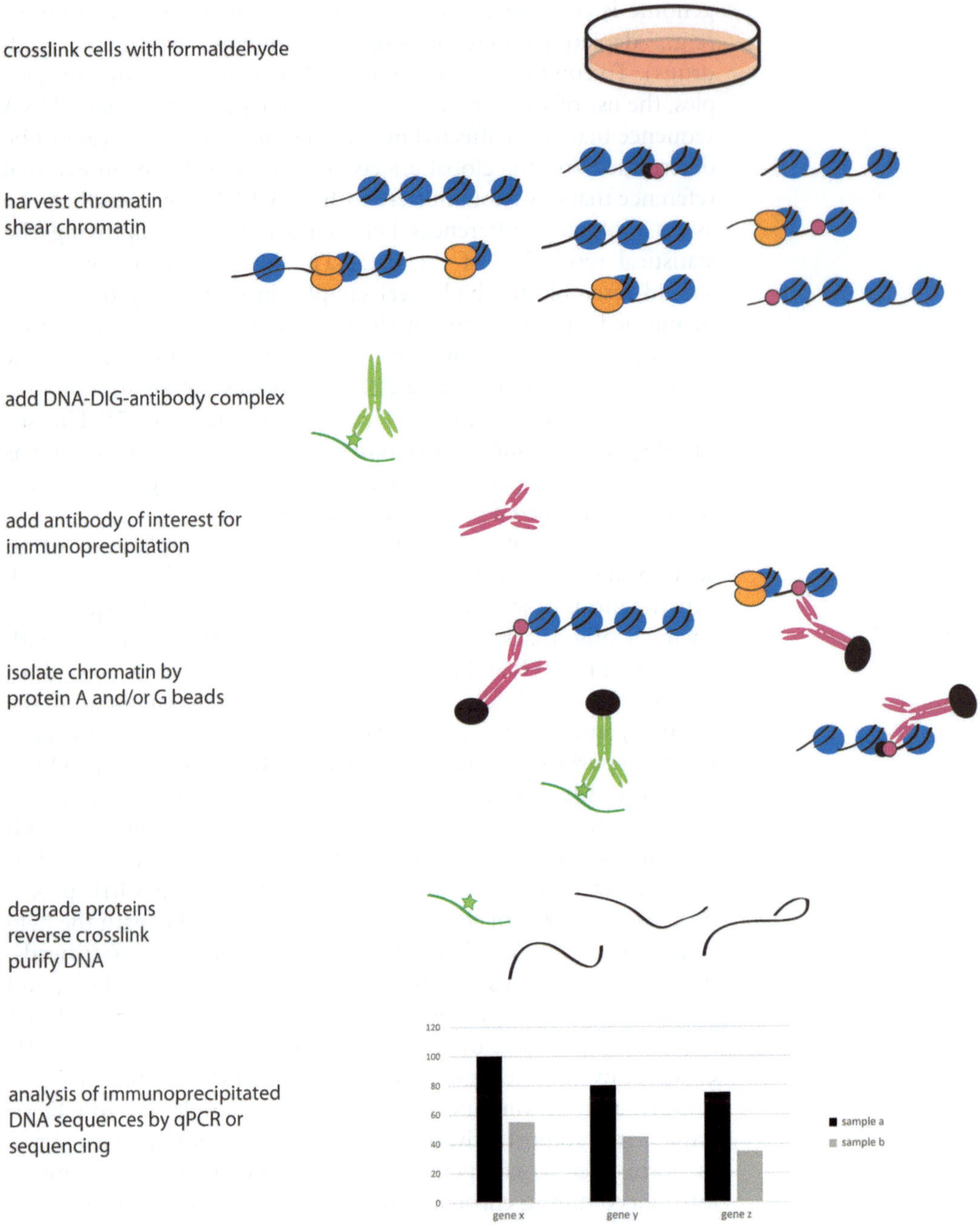

Fig. 2 Schematic overview of the use of an external reference in a ChIP experiment. A typical ChIP experiment starts by treating living cells with formaldehyde to crosslink proteins to DNA. In the next step, cells are harvested, the chromatin extracted and sheared by sonication or enzymatic digestion to obtain DNA fragments of about 500 bp. A constant amount of the DNA–DIG–antibody complex (in *green*) is then added as an external normalizer to each chromatin extract before immunoprecipitation with the antibody of interest (in *purple*). The DNA–DIG–antibody complex is co-isolated with the chromatin bound by the antibody of interest by means of magnetic or sepharose beads covered with protein A and/or G (*black circle*). In the end, proteins and RNA are degraded, the crosslinking reversed and the DNA is purified. The purified DNA sequences are finally quantified by qPCR or sequencing

designed primers for the qPCR do not amplify any sequences in chromatin extracts of HeLa and Drosophila S2 cells. DIG, which is used as an antigen conjugated to DNA, is exclusively found in the plant of the genus digitalis and the antibody does not cross react with any cellular protein in animal cells.

2 Materials

2.1 Production of the DNA Fragment

1. Distilled water.
2. dNTPs stock solution (10 mM each).
3. Taq DNA polymerase (e.g., from Thermo Fisher Scientific).
4. 10× buffer for Taq DNA polymerase.
5. 25 mM $MgCl_2$ (if not included in 10× buffer).
6. DNA template for PCR reaction: e.g., pETcyo plasmid [9].
7. Oligonucleotides for amplification of DNA fragment: Forward oligo: 5′-gtgcgcgaacgtactgatta-3′ and reverse oligo: 5′-agatagcgatccagggtcaa-3′. There is no special requirement for purification of the oligonucleotides.
8. Tubes for PCR.
9. PCR machine.
10. Equipment and reagents for agarose gel electrophorese: agarose, TAE buffer (40 mM Tris, 20 mM acetic acid, 1 mM EDTA), ethidium bromide, loading buffer, molecular-weight size marker.
11. DNA purification kit, e.g., illustra GFX PCR DNA and Gel Band Purification Kit (GE Healthcare Life Sciences).
12. UV spectrophotometer/NanoDrop instrument.

2.2 Digoxigenin Labeling and Production of DNA–DIG–Antibody Complex

1. 100 mM dATP, 100 mM dCTP, 100 mM dGTP, 100 mM dTTP.
2. Digoxigenin-11-2′-deoxy-uridine-5′ triphosphate (DIG-11-dUTP), alkali-stable (Roche).
3. Anti-digoxigenin (DIG) antibody, e.g., mouse monoclonal anti-DIG antibody (Abcam).
4. PBS buffer: 137 mM NaCl, 2.7 mM KCl, 10 mM Na_2HPO_4, 2 mM KH_2PO_4.
5. Rotating wheel.
6. 37% formaldehyde.
7. 1 M glycine (pH not adjusted).
8. Protein concentrator cut off 100 kDa, e.g., Nanosep 100K omega centrifugal devices (Pall Gelman Laboratory).

2.3 DNA–DIG–Antibody Complex in a ChIP-qPCR Experiment

1. Reagents and equipment for ChIP experiment.
2. Reagents and equipment for qPCR: qPCR machine, qPCR tubes, 2× master mix for qPCR, e.g., KAPA SYBR Fast qPCR Master Mix (KAPA Biosystems).
3. Oligonucleotides for qPCR: NP-F: 5′-tattgcttccttcccaattctg-3′ and NP-R: 5′-gtcaacaacgcgacggtaa-3′. There is no special requirement for purification of the oligonucleotides.

3 Methods

3.1 Pre-experiments to Get a Suitable DNA Fragment

Choose a DNA sequence that is not present in the genome of the organism of interest and thus does not interfere in the ChIP experiment. Here, a part of the *bo3* oxidase operon from *E. coli* which is absent in the genome of eukaryotes is used [9]. For choosing alternative DNA sequence *see* **Notes 1** and **2**.

In a first step, the DNA sequence is amplified and purified. To check whether the chosen DNA sequence is suitable as external reference in ChIP experiments, the amplified product is detected with nested primers in a qPCR run.

1. Pipette the following reagents in a PCR tube (total volume is 100 μl; the sequences of the primers are listed in Subheading 2.1 above):

H_2O	76 μl
10× buffer (for Taq DNA polymerase)	10 μl
$MgCl_2$ (25 mM)	6 μl
dNTP (10 mM each)	2.5 μl
pETcyo plasmid (10 ng/μl)	1 μl (*see* **Note 3**)
forward oligo (20 μM)	2 μl
reverse oligo (20 μM)	2 μl
Taq DNA polymerase (5 U/μl)	0.5 μl

2. Run a PCR reaction according to the following scheme:

   1× (95 °C—5 min); 35× (95 °C—30 s, 55 °C—30 s, 72 °C—30 s); 1× (72 °C—7 min).
3. Analyze 5 μl of the PCR reaction on an 1.5% agarose gel (*see* **Note 4**).
4. Purify the PCR product by a PCR purification kit.
5. Run a real-time PCR (quantitative PCR, qPCR) with the purified PCR product as template and nested primers for the DNA fragment (NP-F and NP-R). A typical reaction in a qPCR experiment consists of 10 μl 2× master mix (containing a suitable buffer, $MgCl_2$, SYBR green, dNTPs, and Taq DNA

polymerase, e.g., KAPA SYBR Fast qPCR Master Mix), 3 µl H_2O, 2 µl DNA template, and 5 µl forward and reverse oligonucleotides (NP-F and NP-R, 4.5 µM each, oligonucleotide sequences are listed in Subheading 2.3 above). Use different dilutions of the PCR fragment to monitor the reaction efficiency of the qPCR and analyze the melting curves to confirm the suitability of the nested primers. Include in the same run H_2O instead of DNA as a non-template control. In addition, take genomic DNA of the organism of interest instead of the PCR product as template to ensure that the nested primers do not amplify any sequence of the chromatin in the ChIP experiments (no signal above background should be detected, *see* **Note 5**).

3.2 Labeling of DNA with Digoxigenin by PCR

The chosen and tested (*see* Subheading 3.1) DNA sequence is labeled by PCR using the commercially available DIG-labeled nucleotide digoxigenin-11-2′-deoxy-uridine-5′ triphosphate (DIG-11-dUTP) which is incorporated into newly synthesized DNA instead of dTTPs (*see* **Note 6**).

1. Mix 5 µl dATP (100 mM), 5 µl dCTP (100 mM), 5 µl dGTP (100 mM), 4.5 µl dTTP (100 mM), and 30.5 µl H_2O to prepare a stock of dNTP (final concentration: 10 mM dATP, 10 mM dCTP, 10 mM dGTP, and 9 mM dTTP).
2. Set up a PCR reaction with spiked DIG-11-dUTPs which is integrated into the PCR product (total volume 100 µl; the sequences of the primers are listed above in Subheading 2.1):

H_2O	75 µl
10× buffer (for Taq DNA polymerase)	10 µl
$MgCl_2$ (25 mM)	6 µl
dNTP mix from **step 1**	2 µl
DIG-11-dUTP (1 mM)	1 µl
pETcyo plasmid (10 ng/µl)	1 µl
forward oligo (20 µM)	2 µl
reverse oligo (20 µM)	2 µl
Taq DNA polymerase (5 U/µl)	1 µl

3. Run a PCR reaction according to the following scheme:

   1× (95 °C—5 min); 35× (95 °C—30 s, 55 °C—30 s, 72 °C—30 s); 1× (72 °C—7 min).
4. Purify the PCR reaction by a PCR purification kit to remove nucleotides including DIG-11-dUTPs (*see* **Note 7**).
5. Measure the concentration of the purified PCR product by UV spectrophotometry, e.g., NanoDrop instrument.

3.3 Production of DNA–DIG–Antibody Complex

1. The amount of anti-DIG antibody needed for preparing the DNA–DIG–antibody complex is calculated so that each DIG molecule of the DNA fragment is bound by an antibody, assuming that every IgG antibody recognizes two DIG molecules. Take the formula below to calculate the amount of antibody that is used for the DNA–DIG–antibody complex formation (*see* **Note 8** for clarification):

$$\text{ng (antibody)} = \frac{\text{ng(DNA fragment)} \times \text{kDa (antibody)}}{\text{kDa (DNA fragment)}} \times \frac{\text{DIG/DNA fragment}}{2}.$$

2. Incubate the purified, DIG-labeled DNA fragment (PCR product from Subheading 3.2) with the calculated amount of the anti-DIG antibody in PBS for 2 h on a rotating wheel at room temperature. A typical reaction contains 500 ng DNA fragment and 2000 ng antibody in 100 μl PBS.
3. Add 5.4 μl formaldehyde (37%) to the tube (*see* **Note 9**).
4. Incubate for 10 min at room temperature.
5. Add 12.5 μl 1 M glycine to quench the aldehyde groups and terminate the crosslinking reaction.
6. Incubate for 10 min at room temperature.
7. To remove residual formaldehyde, wash the DNA–DIG–antibody complex with PBS by a spin column with a cut off of 100 kDa (*see* **Note 10**), e.g., Nanosep 100K omega centrifugal devices.
8. Dilute the DNA–DIG–antibody complex in PBS to a final volume of 1 ml.
9. Aliquot the complex (20 μl) and store at −80 °C (*see* **Note 11**).

3.4 Test the Specificity and Working Concentration of DNA–DIG–Antibody Complex

1. Prepare chromatin for a typical ChIP experiment according to your protocol (*see* **Note 12**).
2. Divide the chromatin into five tubes (tube 1–5).
3. Add different amounts of the DNA–DIG–antibody complex to the tubes containing chromatin (tubes 1–3): In an initial test experiment, use 0, 5, and 50 μl of the complex (*see* **Note 13**). As negative controls, add the DNA fragment only and the anti-DIG antibody only into two separate tubes with chromatin (tubes 4 and 5).
4. Incubate for 15 min on a rotating wheel in the cold room.
5. Remove 1/100 of the sample from each tube and transfer to a new tube (inputs 1–5).
6. Add magnetic or sepharose beads that are covered with protein A and/or G to the tubes and incubate for 1 h on a rotating wheel in the cold room (*see* **Note 14**).

7. Wash beads according to your ChIP protocol.
8. Digest the proteins and RNA as well as reverse the crosslink according to your standard ChIP protocol. Include here the input samples from **step 5**.
9. Purify the DNA by phenol–chloroform extraction or by purification columns.
10. Run a qPCR with the purified samples (input and immunoprecipitated material) and use the nested primers NP-F and NP-R to amplify the DNA sequence of the DNA–DIG–antibody complex. A typical reaction in a qPCR experiment consists of 10 μl 2× master mix (containing a suitable buffer, $MgCl_2$, SYBR green, dNTPs, and the Taq DNA polymerase), 3 μl H_2O, 2 μl DNA template, and 5 μl forward and reverse oligonucleotides (NP-F and NP-R, 4.5 μM each, oligonucleotide sequences are listed Subheading 2.3). In the same run, include H_2O instead of DNA as a non-template control.
11. Analyze the threshold cycle values (Ct) of the DNA fragment by setting the input value to 1 and by calculating the immunoprecipitated material relative to the corresponding input sample. No signal should be obtained for the "antibody only control" and for the immunoprecipitated "DNA only control" sample.

3.5 The DNA–DIG–Antibody Complex in a ChIP Experiment

During a ChIP experiment, the DNA–DIG–antibody complex is added to the reaction after shearing of the chromatin and before the incubation with the antibody of interest. It is crucial that this external reference is already present in the input sample (starting material), therefore ensure that the complex is added before the input is removed from the sample. The DNA–DIG–antibody complex does not disturb the ChIP experiment and the entire ChIP procedure remains the same (Fig. 2).

1. Add the DNA–DIG–antibody complex to the chromatin before the addition of the antibody.
2. Follow your standard ChIP protocol.
3. Finally, quantify the DNA sequence in the starting material (input) and in the immunoprecipitated material by the nested oligos NP-F and NP-R in a qPCR run (*see* also **step 10** in Subheading 3.4). These values can be used for normalization of the gene/DNA of interest as external reference and can therefore compensate for variations between the samples (e.g., loss during the purification of DNA) (Fig. 1, *see* **Note 15**).

4 Notes

1. Since the chosen DNA fragment is added to the chromatin extract during the ChIP experiment and quantified by qPCR or sequencing, it is essential that the DNA sequence does not cross-react with any sequence in the genome of the organism of interest. Therefore, homology searches between the chosen DNA sequence and the genome of the favorite organism should be performed (e.g., Blastn).
2. The DNA fragment should have a minimum length of 70 nt to enable robust quantification with nested primers by qPCR.
3. For the PCR, a plasmid containing the quinol *bo3* oxidase (pETcyo) was used. However, it is possible to use any other DNA sequence that does not cross-react with sequences in the genome of your organism of interest.
4. A single DNA band of 78 bp should be present on the agarose gel if using the indicated plasmid and oligonucleotides.
5. To investigate unspecific binding of the oligonucleotides to the genomic DNA of the favorite organism, the DNA should mimic the chromatin extract of a typical ChIP experiment. Therefore, the DNA should be sheared by sonication or by enzymatic digestion as in ChIP experiments. If available, a sample/input from a previous ChIP experiment can be used.
6. The ratio between DIG-11-dUTPs and dTTPs can be varied for different labeling extent of the resulting PCR product. Ensure that at least one DIG-labeled dUTP is incorporated per DNA fragment. Here, a ratio between DIG-11-dUTPs and dTTPs of 1:18 was chosen, i.e., on average, 1.4 dTTPs per ssDNA sequence are exchanged by DIG-11-dUTPs (26 thymidines are present in the 78 nt DNA sequence).
7. A small aliquot of the PCR reaction can be analyzed on an analytical 1.5% agarose gel, as described for **step 3** in Subheading 3.1.
8. The amount of the DNA fragment (ng) is measured by a UV spectrophotometer (*see* **step 4**; Subheading 3.2). The molecular weight of an IgG antibody is approximately 150 kDa and the molecular weight of the DNA fragment is estimated by multiplying the number of base pairs with 650 Da, the average molecular weight of a base pair (here: $78\ \text{bp} \times 650\ \text{Da} = 50.7\ \text{kDa}$). In addition, the formula takes into consideration the amount of incorporated DIG-labeled dUTP per DNA fragment (here, since the DNA fragment is double-stranded: $2 \times 1.4 = 2.8$) and the two binding sites for DIG per antibody.

9. Crosslinking was performed with 2% formaldehyde. The crosslinking can be modulated by the amount of formaldehyde or by other crosslinking reagents. It is reasonable to use the same crosslinking conditions as for the chromatin in your standard ChIP experiment.
10. Using a concentrating unit with a cut off of 100K, DNA fragments that are not bound by an antibody will be washed out, decreasing background signals.
11. Analysis of the stability of the DNA–DIG–antibody complex showed that the complex must be stored at −80 °C or at −20 °C in the presence of glycerol [8]. It is therefore essential to prepare small aliquots of the complex and to avoid repeated freezing–thawing cycles.
12. Treat the chromatin extracts according to your standard ChIP protocol: pre-clear the samples if necessary and use the buffer including detergents that is required for the antibody incubation.
13. The signals of the normalizer DNA fragment measured by qPCR should be in the same range as the signals for the gene/DNA of interest. Therefore, depending on the experiment, the amount of the DNA–DIG–antibody complex has to be adjusted.
14. Take the same amount of beads and incubation time as in your standard ChIP protocol.
15. If the ChIP experiment is analyzed by high-throughput sequencing (ChIP-seq), the DNA fragment will also be sequenced and can therefore be used as reference to compensate for variability between samples.

Acknowledgment

I am grateful to Profs Neus Visa (Stockholm University) and Christoph von Ballmoos (University of Bern) for critical reading of the manuscript.

References

1. Aparicio O, Geisberg JV, Struhl K (2004) Chromatin immunoprecipitation for determining the association of proteins with specific genomic sequences in vivo. Curr Protoc Cell Biol 17:7. doi:10.1002/0471143030.cb1707s23
2. Gordân R, Hartemink AJ, Bulyk ML (2009) Distinguishing direct versus indirect transcription factor-DNA interactions. Genome Res 19 (11):2090–2100
3. Listerman I, Sapra AK, Neugebauer KM (2006) Cotranscriptional coupling of splicing factor recruitment and precursor messenger RNA splicing in mammalian cells. Nat Struct Mol Biol 13(9):815–822
4. Hoffman EA, Frey BL, Smith LM et al (2015) Formaldehyde crosslinking: a tool for the study of chromatin complexes. J Biol Chem 290 (44):26404–26411

5. Goens G, Rusu D, Bultot L et al (2009) Characterization and quality control of antibodies used in ChIP-assays. Methods Mol Biol 567:27–43
6. Peach SE, Rudomin EL, Udeshi ND et al (2012) Quantitative assessment of chromatin immunoprecipitation grade antibodies directed against histone modifications reveals patterns of co-occurring marks on histone protein molecules. Mol Cell Proteomics 11(5):128–137
7. Bonhoure N, Bounova G, Bernasconi D et al (2014) Quantifying ChIP-seq data: a spiking method providing an internal reference for sample-to-sample normalization. Genome Res 24(7):1157–1168
8. Eberle AB, Böhm S, Östlund Farrants AK et al (2012) The use of a synthetic DNA-antibody complex as external reference for chromatin immunoprecipitation. Anal Biochem 426 (2):147–152
9. Frericks HL, Zhou DH, Yap LL et al (2006) Magic-angle spinning solid-state NMR of a 144 kDa membrane protein complex: E. coli cytochrome bo3 oxidase. J Biomol NMR 36 (1):55–71

INDEX

Neus Visa and Antonio Jordán-Pla (eds.), *Chromatin Immunoprecipitation: Methods and Protocols*,
Methods in Molecular Biology, vol. 1689, DOI 10.1007/978-1-4939-7380-4, © Springer Science+Business Media LLC 2018

G

H

I

L

M

N

MIX
Papier aus verantwortungsvollen Quellen
Paper from responsible sources
FSC® C105338

If you have any concerns about our products,
you can contact us on
ProductSafety@springernature.com

In case Publisher is established outside the EU,
the EU authorized representative is:
Springer Nature Customer Service Center GmbH
Europaplatz 3, 69115 Heidelberg, Germany

Printed by Libri Plureos GmbH
in Hamburg, Germany